BIOCHEMISTRY

Contents

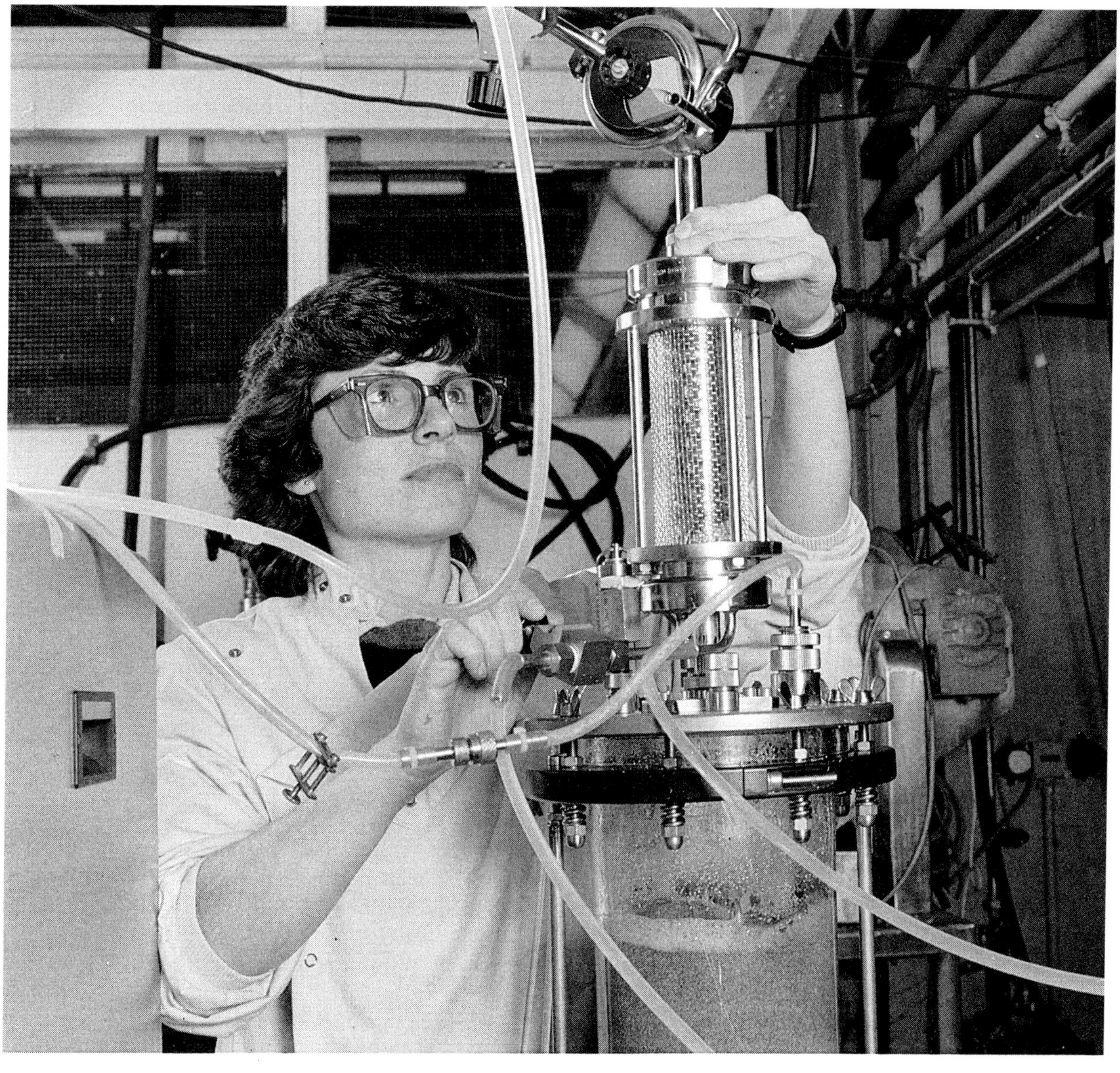

A biochemical engineer testing equipment designed for use in the manufacture of biochemicals.

CHAPTER 1 INTRODUCING BIOCHEMISTRY

1.1 Biochemistry

Biochemistry, as its name suggests, is the area where the interests of biologists and chemists meet. Biochemists seek an understanding of the chemical structure of living organisms and the chemical processes which take place in them. They then hope to put this knowledge to good use.

Biochemistry has grown from the study of individual compounds, such as hormones, to the study of the networks of reactions involved in processes like photosynthesis. Today biochemists study how chemical processes take place in cells, how these processes are controlled and how chemical compounds are used to pass on information from one generation of living organisms to the next.

Most of the biochemistry in this book dates from only the last 50 years, whereas the chemistry you are studying was developed over the last two centuries. An example is the work of the double Nobel prize winner Frederick Sanger who in 1955 described the very first sequencing of the amino acids in a protein, insulin, and then in 1977 published the full sequence of bases in the DNA of a virus. In 1997 he was the guest of honour at the opening of new laboratories concerned with the complete sequencing of human DNA, the human genome project.

1.2 Experimental work in biochemistry

When you neutralize 25 cm^3 of 1.0 M hydrochloric acid with 1.0 M sodium hydroxide in the laboratory, the volume of sodium hydroxide required will always be 25 cm^3. Furthermore the mass of sodium chloride obtained, if the solution is allowed to evaporate to dryness, will always be the same. The same experiment can be repeated in any laboratory in the world and the same results obtained.

You might expect the same to be true for biochemistry experiments. However, there are often large differences between the results of a biochemistry experiment apparently carried out in identical ways. You will probably find during this Special Study that different groups doing the same experiment may get very different results.

Why should this be? The chemical reactions which a biochemist studies are no different from the reactions that an organic chemist studies. However, biochemical reactions are catalysed by enzymes, which are very sensitive to the smallest change in conditions. Furthermore biochemists often have to start their experiments by obtaining an enzyme from living cells. The amount of enzyme obtained will depend on how much was present in the original cells, and that will certainly vary from one source to another.

For example, the activity of your enzyme preparations in the experiments in Chapter 2 will depend on how effectively you carry out the extraction process. Insufficient breaking up of the cells will result in little enzyme being released and most of the enzymes will still be locked away within intact cells. Furthermore, any delays in using your extract will result in the extract having reduced catalytic activity. So you must expect a range of results in your practical work and on occasions an experiment may not work!

COMMENT

Biochemists have to measure very small biological structures and they often use micrometres for sizes and microlitres for volumes. A micrometre is a millionth of a metre (1 μm = 10^{-6} m). Atoms of course are even smaller and usually measured in nanometres (1 nm = 10^{-9} m).

Example	Diameter / μm
Hydrogen atom	0.00007
Carbon atom	0.0002
Lysozyme enzyme	0.003
Polio virus	0.03
Cell nucleus	3
Human cell	50

But in contrast the total unravelled length of DNA in a human cell is 2 metres!

1.3 Biochemistry and medicine

The relationship of biochemistry to medicine has always been close: advances in biochemistry result in new insights about our health, such as a better understanding of nutrition.

Although the causes are various, all diseases can be regarded as a breakdown in the normal chemical processes in our bodies. Doctors treated illnesses with plant extracts long before the active ingredients had been identified: an example is the treatment of a fever with willow bark, now replaced by aspirin. However, with an understanding of how a disease alters the chemistry of affected cells it becomes possible to 'design' molecules that are likely to correct a condition.

Another area where biochemistry has an important role is in the diagnosis of illness. Doctors can deduce a lot by taking your temperature, looking at your throat and feeling for painful places under your skin. But nowadays you are often asked to provide samples of blood and urine that will be put through a set of biochemical tests to help decide what is wrong with you. Biochemical tests can reveal which chemical processes in your body are not working properly: doctors use test sticks to indicate the glucose level in urine as a preliminary test for diabetes and diabetics can monitor their blood glucose levels in a similar way (see figure 1.1). For some babies the importance of

NUFFIELD ADVANCED SCIENCE

CHEMISTRY

BIOCHEMISTRY

A Special Study

Published for the Nuffield Foundation by Addison Wesley Longman
Third edition

Editor of this Special Study
Michael Vokins

Author of this Special Study
Roger Barnes

Contributors
Sue Crickmore
Ray Mathias (John Innes Centre)
Richard Price (SAPS)
Michael Reiss
Michael Vokins

Safety adviser
Peter Borrows

Acknowledgement is due to those who helped with the original development of the *Biochemistry* Special Study.

The text published in 1970 was developed by a team drawn from schools, universities and industry, and edited by Gordon Raitt.

The revised text published in 1984 was written by Roger Barnes and Roger Norris, editor Andrew Hunt.

Drawings by Hugh Neill, Gary Simmons, Ian Smith, and Abigail Dodd

Structure diagrams by Speedlith Limited

Layout by Stephen Pargeter

Photographs:
Cover: Zeneca
Contents page: BEES, University College, London
Action Images: fig 3.1
Cavendish Laboratory, University of Cambridge: fig 4.6 below
Cellmark Diagnostics, Abingdon, UK: fig 4.17
John Chadwick, PPL Therapeutics: fig 5.4
Don Helinski, University of California, San Diego: fig 5.2
Hulton Getty Images: fig 5.1 above
Oscar J Miller, University of Virginia: fig 4.16
Natural History Museum: fig 1.2
Natural Portrait Gallery: fig 4.6 above
Science Photo Library: fig 1.1, 2.4, 2.5, 4.7, 4.10, 4.12, 5.1 below, 5.5

Picture research: Catherine Blackie

Editorial: Donna Evans and Sarah Codrington

The Nuffield Foundation and the Publishers are grateful to Edexcel Foundation for permission to produce revised versions of answers to questions from past examination papers.

The publishers would like to thank the following organisations for permission to base illustrations in this book on their publications:

page 9, P T Khaw and A R Elkington, *ABC of eyes*, BMJ
page 15, National Dairy Council
page 28, T M Devlin, *Biochemistry*, John Wiley
page 27, 29, 31, 34, D Micklos and G Freyer, *DNA Science*, Carolina Biological Supply Company
page 25, Alberts, Bray, Lewis, Raff, Roberts and Watson, *Molecular Biology of the Cell*, Garland Publishing Inc
page 26, 30, E J Wood, C A Smith and W R Pickering, *Life Chemistry and Molecular Biology*, Portland Press
page 34, Experiment 4.6 adapted from material by SAPS

Pearson Education Limited

Edinburgh Gate, Harlow,
Essex, CM20 2JE and Associated Companies throughout the world

First published 1970
Second impression 1999

This edition 1998

Set in Times and Univers

Printed in Singapore (FOP)

ISBN 0 582 31188 8

biochemical testing cannot be overestimated because life-threatening conditions can be identified even before the symptoms appear – and for some conditions, such as phenylketonuria, that would be too late.

In the following chapters you will find some MEDICAL CASE STUDIES which illustrate the relationship of biochemistry to medicine.

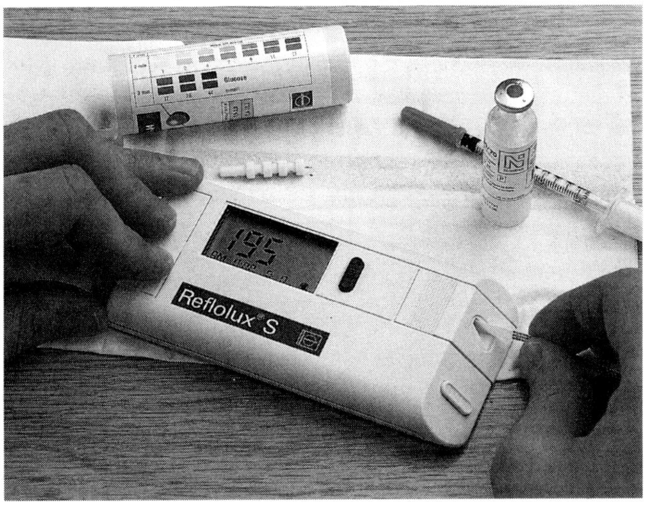

Figure 1.1 The colour reaction on an enzyme-based test stick can be measured with a specialized colorimeter to determine blood glucose levels. The insulin is obtained from genetically engineered sources.

1.4 Molecular biology

The most recent benefits from biochemistry are related to the spectacular advances in our knowledge of DNA and the development of the branch of biochemistry called molecular biology. Molecular biology is concerned with the study of cells at the molecular level.

The origins of many inherited conditions have been traced to faults in the chemical composition of DNA and this has lead to a much greater understanding of the nature of the condition. However, understanding the cause does not automatically lead to a suitable treatment: people suffering from an inherited condition can often only be offered better treatment at present, not a cure.

Whole new industries have sprung up based on DNA technologies. Useful characteristics can be transferred from one species to another by transferring pieces of DNA, creating what are known as transgenic organisms. The products obtained from such organisms can be as important as insulin, or as apparently frivolous as a naturally blue cotton for making jeans. When the products are proteins like insulin the manufacturing processes require a new approach to the design of the equipment. If the wrong technique is used to stir the reaction mixture the protein molecule may be affected and become inactive; rather like shaking milk can produce lumps of cream.

Another area where DNA techniques are making a contribution is in our understanding of evolution. The closer species are in their evolutionary history the more similar their DNA is; the further apart the more dissimilar their DNA. Some questions about evolution have been answered by DNA studies. In 1997 German scientists re-examined the bones of a Neanderthal man (see figure 1.2), who lived in the Neander valley 30 000 to 100 000 years ago. They found that DNA extracted from the bones had so many differences from our DNA that we are most unlikely to have descended from the Neanderthals.

However, the advance in DNA technology raises important issues. There is concern about the impact on the way we live, about the dangers of genetic engineering as well as the benefits. We shall return to this point in the last chapter after you have learnt more about biochemistry.

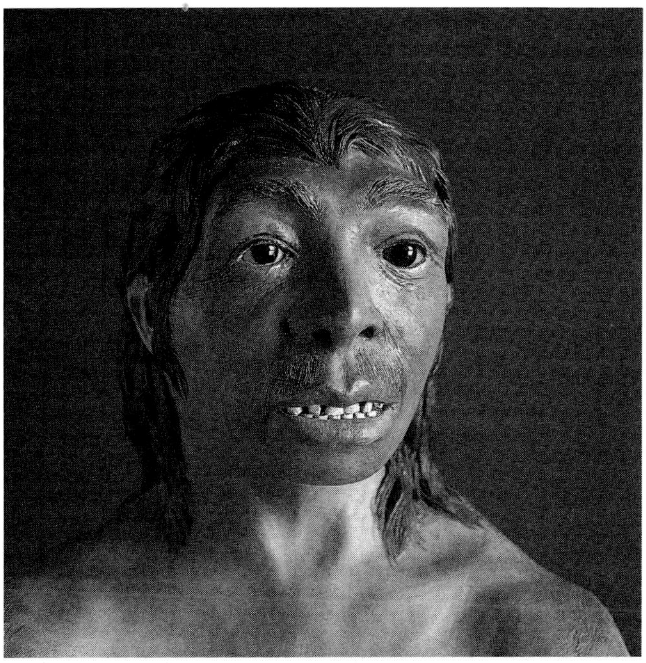

Figure 1.2 Model of a Neanderthal woman: DNA tests suggest that we are not descended from Neanderthals.

REVIEW TASK

This book is not a complete guide to all the chemistry you need to know in order to study biochemistry. You will find it useful to look again at some of the Topics in your main chemistry course that are closely linked to biochemistry, in particular:

Intermolecular forces
Rates of reaction
Natural products

Some of the Study tasks provide additional background material:

the manufacture of citric acid (12.5); the chemical and structural investigation of proteins (14.5); diabetes (Reading task 4); drugs (Reading task 5).

CHAPTER 2 ENZYMES

2.1 The structure of cells

The cell is the fundamental unit of life. Most living organisms are composed of cells which contain nuclei, in which the genetic material is stored. There are single cell organisms, such as amoeba, but the familiar plants and animals which we see every day are built up of many millions of cells. Within each such organism there are many types of cells, with different chemical compositions, structures, and functions. Muscle cells contain contractile fibres, the red blood cell contains haemoglobin to transport oxygen.

However, despite the apparent diversity, all plant and animal cells have a remarkably similar basic composition and structure, and many of the chemical changes which take place in the cells are essentially the same. For instance, the pathways by which glucose molecules are broken down to give energy are virtually the same in both a yeast cell and in a human muscle cell. Minor differences do exist, but they are not so striking as the similarities.

Living cells have all been found to contain inorganic ions such as potassium, chloride and phosphate ions and small organic molecules such as pyruvic acid and glycerol (see box below). A major component of the cell content, however, is very large molecules with molar masses from tens of thousands to several million. Among the most important of these are the **enzymes**, nearly all of which are proteins, and the **nucleic acids** (see Chapter 4). The enzymes are responsible for the catalysis of all chemical reactions in the cell. The nucleic acids contain the information which controls the synthesis of all proteins and are the physical means of passing this information from one generation to the next.

NAMING BIOCHEMICALS

Pyruvic acid is the trivial name for the acid, 2-oxopropanoic acid, and glycerol is the trivial name for the alcohol, propane-1,2,3-triol. Biochemists prefer trivial names to systematic chemical names because of the complexity of many of the molecules met in biochemistry.

$CH_3–CO–CO_2H$	$CH_2OH–CHOH–CH_2OH$
2-oxopropanoic acid	propane-1,2,3-triol

Many biochemicals contain carboxylic acid groups. They are therefore weak acids but at the pH of our bodies they exist as their anions. For instance, in cells at a pH of about 7.2 pyruvic acid exists as the pyruvate anion and is called **pyruvate**.

$$CH_3COCO_2H \rightleftharpoons CH_3COCO_2^- + H^+ \quad K_a = 3.2 \times 10^{-3}\ \text{mol dm}^{-3}$$

pyruvic acid	pyruvate anion
weak acid	conjugate base

The cells of both plants and animals (figure 2.1) are all bounded by a **membrane**, and they contain a variety of other structures which are surrounded by a fluid, the **cytoplasm**. Most of the structures inside a cell have definite functions which are the same for both animal and plant cells. Plant cells have, in addition, a cell wall surrounding the cell membrane and they often contain chloroplasts, the site of photosynthesis.

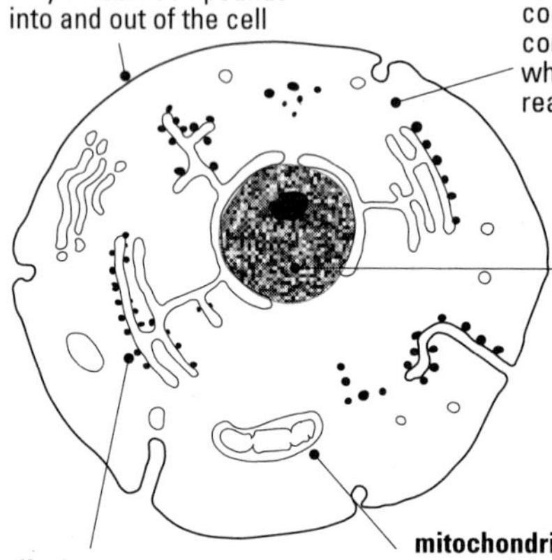

Figure 2.1 General features of animal cells.

Extracting enzymes from cells

Biochemists depend on the techniques used in chemistry, for separating organic materials, for analysis, and for measuring rates of reaction. By studying the reactions which take place in cells biochemists are building an understanding of how cells function. But during biochemical experiments, the delicate nature of living material has to be constantly born in mind. Like any other scientist, a biochemist will want to produce experimental results that can be reproduced by other research groups. The following example highlights the considerable problems facing biochemists separating mitochondria to study an enzyme.

Selecting a suitable material to study

The activity of a particular enzyme will depend on many factors; it is therefore necessary for a biochemist to decide precisely which biological material will be used in every experiment. They must first select a suitable species of plant or animal for study and they must then decide which tissue or organ to study. Animal material is only used when there is no other alternative. Samples must be kept cool in a buffer solution to stop the contents of the cells breaking down.

Disrupting the cell membranes

The biochemist must gently break open the cells to release their contents, in a process called homogenization (figure 2.2). This must be done without damaging the cells' contents. Cells differ greatly in the strength of their cell membranes and cell walls.

Yeast cells for instance have a hard cell wall which is very difficult to break. Plant cells also have strong outer walls; they may be broken open using a sharp rapidly rotating metal blade such as the liquidizer attachment of a food processor.

Homogenization produces a mixture of undamaged cells, membranes, mitochondria, ribosomes, cytoplasm, and nuclei which is known as a **homogenate**, even though it is far from being homogeneous.

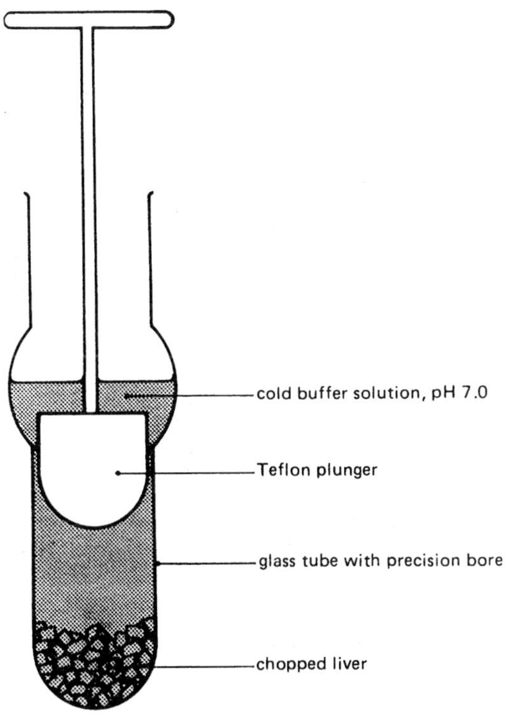

Figure 2.2 The homogenization of cells. The diameter of the tube is only 0.025 cm larger than that of the plunger. Liver cells are broken open as they pass between the plunger and the glass wall but mitochondria are small enough to pass through undamaged.

Separating the cell contents

Now the homogenate is centrifuged in several stages (figure 2.3). Very large centrifugal forces (g) are needed to separate the smallest structures, but refrigerated centrifuges, capable of spinning large volumes of material at high speed for prolonged periods, are available in biochemistry laboratories.

Isolating a pure enzyme

While many enzyme studies may be done with crude homogenates, biochemists often wish to isolate pure enzymes for use in industry or medicine. Various chromatographic techniques can be used in the purification. An example is column chromatography on a gel, which separates proteins according to their size. Usually no one method on its own produces a pure protein; a combination of techniques is necessary if a protein is to be freed from all contaminating material.

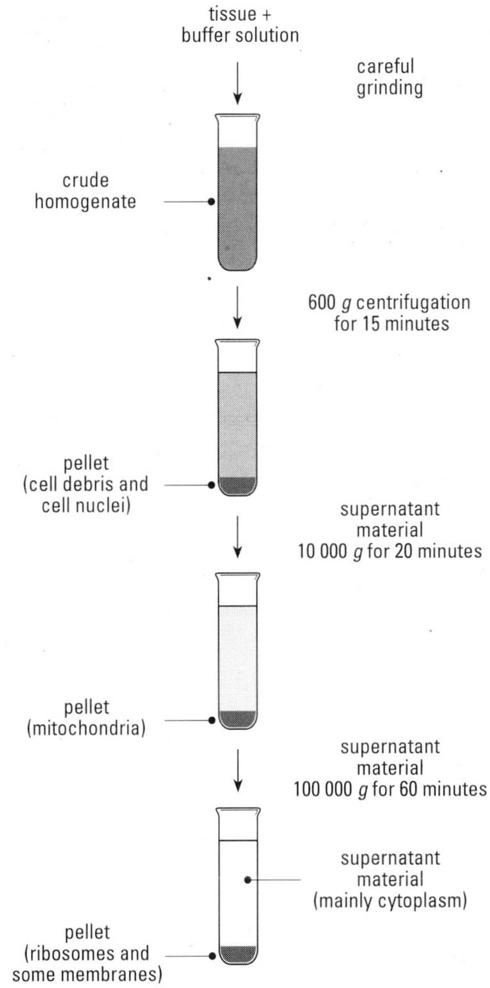

Figure 2.3 Stages in the separation of liver cells' constituents.

Figure 2.4 Some enzymes are produced on an industrial scale. This is a fermentation process and the enzyme is separated by a series of centrifugation and filtration processes.

2.2 Enzymes: proteins as catalysts

Proteins have a critical role to play in nearly all biochemical processes. They can form part of the cellular structure, for example collagen in bones, and they can act as a store of food, for example albumen in eggs. They can be used to transport small molecules around the body, as haemoglobin does in our blood. Complex processes, like muscle contraction, all depend on specific protein molecules. Most important of all, protein molecules act as enzymes, catalysing the thousands of chemical reactions that take place in a single cell.

Biochemists often talk about **enzyme activity.** Enzymes are catalysts and they speed up otherwise slow reactions. Enzyme activity is simply the rate at which a chemical reaction takes place in the presence of an enzyme; it is therefore measured in terms of the rate at which the reactant disappears or product appears, often in mol min^{-1}.

The term **substrate** is used to describe any reactant in an enzyme catalysed reaction. The substrate for catalase (experiment 2.4b) is hydrogen peroxide.

The structure of enzymes

The structure for all proteins is described in terms of their primary, secondary and tertiary structure.

- The **primary structure** is the order in which amino acids are linked together in a protein. You should be able to recall that amino acids are linked by **peptide groups**.

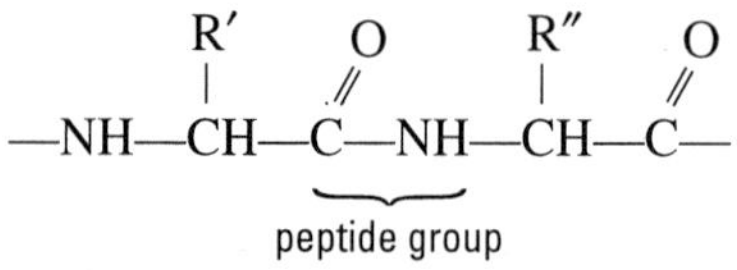

- The **secondary structure** is a description of any regions of the protein where there is regular folding of the protein chain, maintained by hydrogen bonding between peptide groups, such as in the α-helix.
- The **tertiary structure** of the molecule, is a description of the overall three-dimensional shape of the protein. Proteins have complicated but precise shapes maintained by bonds within and between the protein chains.

Many amino acid side chains can form hydrogen bonds. But it is not just hydrogen bonds which maintain the structure of a protein. The carboxylic side chains of some amino acid side chains and the amine groups of others will have lost or gained protons at pH 7, forming negative and positive ions. These can form strong ionic bonds, although there are only relatively few of them in most proteins. Covalent bonds, known as disulphide bridges, —S—S—, can form when two cysteine amino acid side chains react. There are also van der Waals interactions between the large non-polar side chains, which often contain benzene rings. Although the individual van der Waals interactions are weak they make the greatest overall contribution to the stability of the three-dimensional shape of proteins.

Figure 2.5 Fred Sanger, the Cambridge biochemist, was awarded the Nobel prize for the first determination of the primary structure of a protein, insulin.

Let us take the structure of lysozyme as a typical example of enzyme structure. Lysozyme is a water-soluble enzyme present in tears, egg white, and nasal mucus. It has an important role in protecting us from bacteria.

The enzyme is a compact molecule, with dimensions 4.5 × 3 × 3 nanometres. Look at figure 2.7 and you will see that the protein chain starts with an amine group and finishes with a carboxylic acid group. You should also be able to identify three regions of α-helix where the protein chain is folded in a regular way; but most of the chain has an irregular arrangement.

In general, polar side groups lie on the surface of the molecule, where they can interact with water, while most of the non-polar side groups lie within the molecule. Some polar side groups do lie within the molecule and these seem to have a critical role to play in catalysis. Lysozyme acts as a catalyst for the hydrolysis of the carbohydrate cell walls of bacteria. The enzyme has an elongated cleft on the surface, into which a

Amino acid	Side chain	Side chain	Amino acid	Bond between side chains
Alanine	Ⓐ—CH_3	$(CH_3)_2$>CH—Ⓐ	Valine	van der Waals
Serine	Ⓐ—CH_2OH	NH_2COCH_2—Ⓐ	Asparagine	Hydrogen
Aspartic acid	Ⓐ—$CH_2CO_2^-$	$^+NH_3(CH_2)_4$—Ⓐ	Lysine	Ionic
Cysteine	Ⓐ—CH_2—S—S—CH_2—Ⓐ		Cysteine	Covalent

Figure 2.6 The bonds that can occur between amino acid side chains. Ⓐ represents the NH_2—CH—CO_2H group.

chain of six sugars can be fitted (shown in shadow outline in the diagram). This is known as the active site. This type of structure is typical of many enzymes.

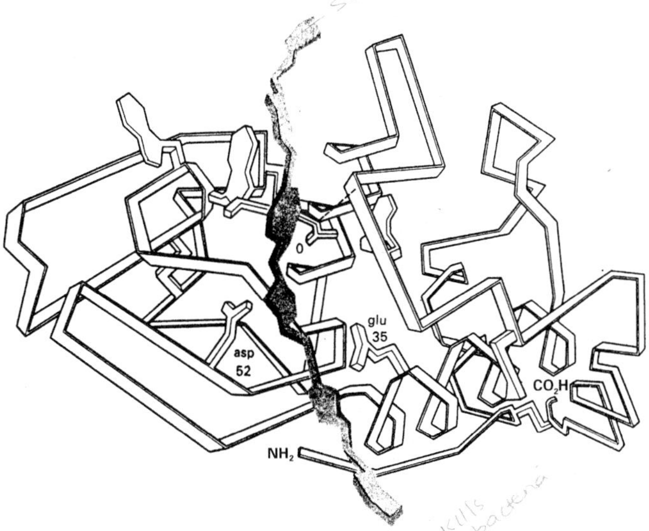

Figure 2.7 The three-dimensional structure of lysozyme, with a carbohydrate molecule shown in the active site. The protein chain starts with an amine group and finishes with a carboxylic acid group; you should also be able to identify three regions of α-helix.

Generalized mechanism of enzyme action

Enzymes are able to catalyse reactions in aqueous solutions under exceptionally mild conditions of temperature and pH. The range of reactions catalysed by enzymes is quite remarkable, for if the chemist confined experiments to such conditions the accomplishments would be very limited. How do enzymes achieve these effects?

The complicated folding of the protein chain in enzymes gives rise to a binding site which is complementary in shape to the substrate. This region is described as the **active site** and it is generally a cleft or crevice on the surface of the enzyme. The site is lined with specific groups which can form weak interactions with particular atoms or groups of the substrate molecule, thus holding it correctly in place. In many cases the presence of the substrate in the binding site induces a small but significant change in the three-dimensional shape of the enzyme. This change in shape results in specific amino acid side chains being brought up to the region of the substrate molecule where the reaction is to take place. Amino acids in the active site may act as proton donors or acceptors, or they may serve to provide a suitable environment for the reaction, either polar or non-polar.

Once the reaction is complete the products diffuse from the active site and the enzyme returns to its original shape.

MEDICAL CASE STUDY The treatment of cataracts in the eye

Difficulty in reading and seeing in bright light can be due to the lens of the eye becoming opaque as a 'cataract' develops. Cataracts are common in people over 60.

The usual treatment is to remove the content of the lens and fit a plastic implant in the lens capsule. Sometimes the complete lens has to be removed and in such cases the enzyme chymotrypsin is used to dissolve the ligaments which hold the lens in place.

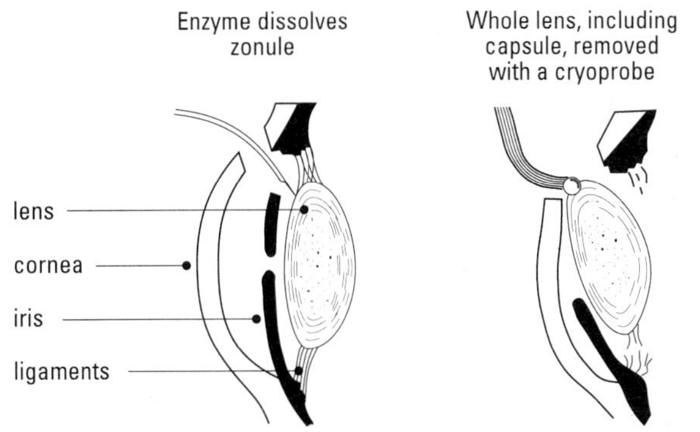

A dilute solution of chymotrypsin is added to the eye cavity and the enzyme dissolves the lens ligaments in a few minutes. The lens can then be sucked out of its cavity. The eye has to be thoroughly washed to avoid damage by the enzyme to other delicate tissues in the eye.

This treatment is much more successful than using surgery to cut the ligaments.

2.3 Factors that affect enzyme activity

Effect of pH

Enzymes, in contrast to many other catalysts, are very sensitive to small changes in pH. Enzymes contain groups that can act as acids and bases, donating or accepting protons, when the pH changes. For example as the pH decreases, more and more of the negatively charged carboxylate groups, $—CO_2^-$, act as bases and accept protons. There will be fewer and fewer negative ions in the enzyme.

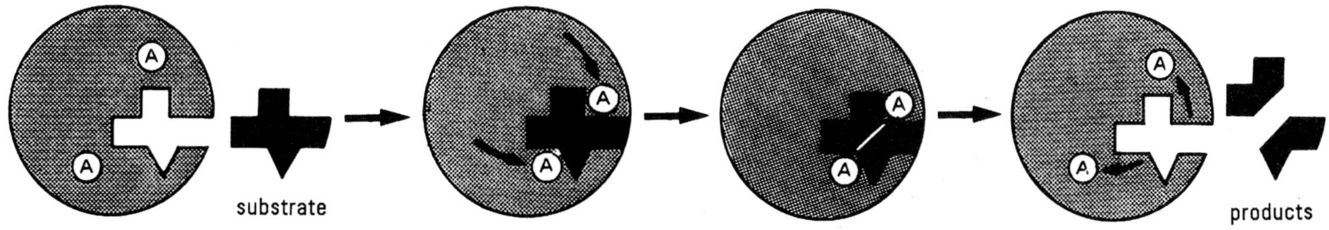

Figure 2.8 The mechanism of enzyme action. Ⓐ represents an amino acid required for catalysis. The binding of the substrate results in a change in the shape of the enzyme and critical amino acids are brought close to the substrate.

> **STUDY TASK Lysozyme**
>
> The active site of the enzyme lysozyme contains two carboxylic acid side chains. It is known that during catalysis one of these acid groups has to be in present as an undissociated acid while the other is needed as a carboxylate ion. The enzyme has a pH optimum between 6 and 7.
>
> 1 Explain what effect increasing the pH to 8 is likely to have on the two carboxylic acid groups and on the activity of the enzyme.
>
> 2 Suggest why the enzyme is inactive at pH 3.

The change in the number of charged groups in the protein affects the number of ionic bonds and hydrogen bonds present. A precise shape is required for maximum enzyme activity and any change in shape is likely to make the enzyme less active.

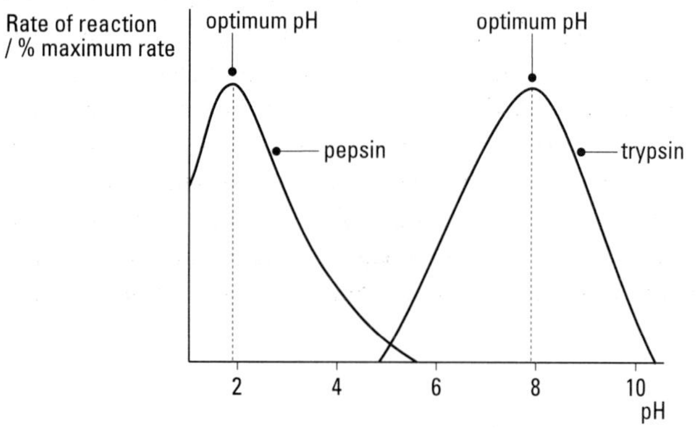

Figure 2.9 The effect of pH on the rate of two enzyme catalysed reactions.

Sometimes acidic or basic groups are specifically required for catalysis. You should recall that acids can catalyse the hydrolysis of esters. Carboxylic acid side chains in an enzyme can catalyse reactions in a similar way by acting as proton donors. However, an increase in pH may result in the loss of the proton from the critical amino acid side chain. If the proton has been lost, the enzyme can no longer offer an alternative pathway for the reaction and it has become inactive.

The optimum pH value for most enzymes lies between pH 5 and pH 9, but there are some exceptions. For instance, pepsin, a enzyme secreted in the stomach, has a pH optimum of about 2. This ensures maximum activity in the acid conditions of the stomach.

Effect of temperature

Temperature has two effects on the rate of an enzyme catalysed reaction. An increase in temperature always increases the number of collisions which have sufficient energy to bring about the reaction. Initially the rate of the reaction increases with increasing temperature until a maximum rate is achieved. However, beyond this temperature the rate of reaction decreases, often dramatically, and this loss of activity is often irreversible.

The activity of an enzyme depends on it maintaining a precise shape and many of the forces maintaining that structure are extremely weak. As the temperature rises, the weak bonds between the chains will break as molecular movement increases and at quite moderate temperatures the molecule will start to unravel and become disordered. On cooling the molecular chains are most unlikely to reform their original highly ordered three-dimensional shapes. When a protein loses its essential three-dimensional shape it has been **denatured**.

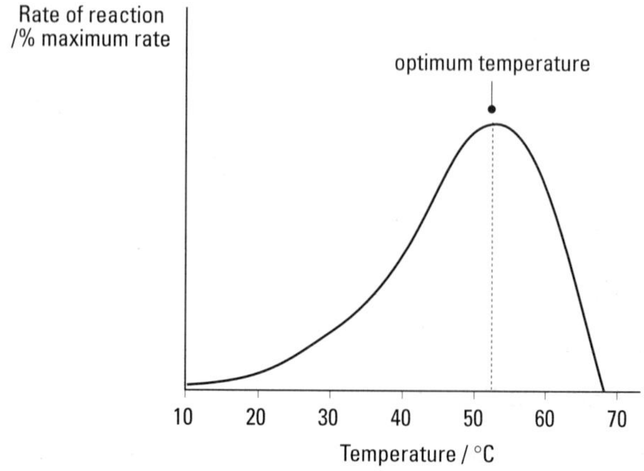

Figure 2.10 The effect of temperature on the rate of an enzyme catalysed reaction.

Effect of substrate concentration

You should use the results of experiment 2.4b to write your own notes on the influence of substrate concentration on enzyme activity.

> **Comment**
>
> Some enzymes are much more susceptible to temperature change than others and each enzyme is said to have a temperature optimum at which it works best. There are microbes which live in the hot springs of the Yellowstone National Park; one micro-organism has been isolated which grows best at about 105 °C and can survive in superheated water at 113 °C! Enzymes which have unusually high optimum temperatures are being used increasingly in industry. One example is Taq polymerase which can be used at 60–90 °C to increase the amount of DNA available for DNA fingerprinting.
>
> By contrast other bacteria live in Antarctic ocean water and one species grows best at only 4 °C, and cannot reproduce if the temperature is above 12 °C.

Specificity

Many enzymes will only act on one specific compound and will only catalyse one type of reaction. The active site has a shape which is complementary to that of the substrate. Furthermore specific bonds will hold the substrate in place. So only molecules of the correct shape and with the correct groups for binding can act as substrates. Activity may even be limited to one particular optical isomer: the enzyme which oxidizes lactate to pyruvate is completely specific for the L optical isomer.

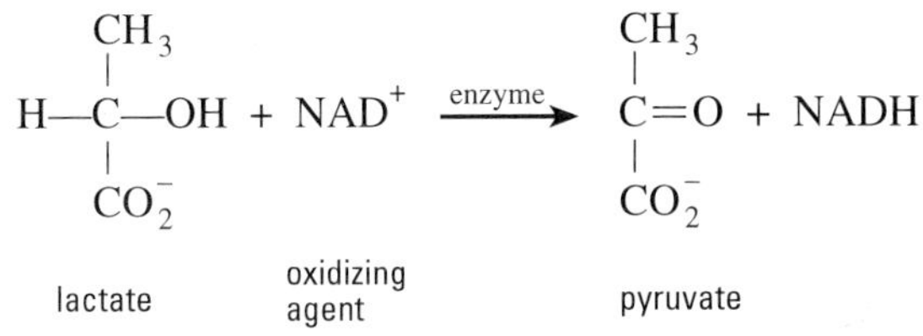

Which is the chiral centre?

STUDY TASK UREASE

Urease will catalyse the hydrolysis of urea but it cannot catalyse similar reactions involving methylurea, ethanamide or thiourea.

1 Look up the formulae of the four possible substrates and draw their displayed formulae. Suggest why urease does not act on three of them.
2 Suggest what features a suitable substrate must have.

Cofactors

Most enzymes have been found to contain non-protein groups, which are essential for catalytic activity. Catalase, an enzyme from liver, contains haem (see figure 3.8a) and in the absence of bound haem, the enzyme is unable to decompose hydrogen peroxide. A number of enzymes contain zinc(II), iron(II) or (III), or copper(II) ions.

STUDY TASK COFACTORS

1 What type of reaction might you expect an enzyme containing iron or copper ions to catalyse?

2 Is the zinc ion likely to fulfil the same function as the copper and iron ions? Explain your answer.

Inhibition

Enzymes are inactivated by any chemical which brings about the disruption of their three-dimensional shape. Strong acids and alkalis or detergents denature proteins and therefore affect enzyme activity. But the activity of many enzymes may also be reduced by specific compounds which have no denaturing effect on proteins. Such chemicals are described as **inhibitors**. The inhibition may be reversible or irreversible depending on whether or not the inhibitor becomes strongly bonded to the protein.

Irreversible inhibitors often bind covalently to the enzyme. They disrupt the essential three-dimensional shape of the enzyme or react with an amino acid side chain which is needed for catalysis. They can be considered as 'enzyme poisons' and include metal ions such as silver and mercury and drugs designed to permanently block the action of an enzyme. For example, the drug deprenyl is used in the treatment of Parkinson's disease. The drug molecule permanently binds by covalent bonds to an enzyme in the brain and blocks its action.

Competitive inhibitors are compounds which are similar in structure to an enzyme's substrate. They bind to the active site of the enzyme and they remain there unchanged. If, however, the concentration of the substrate is greatly increased the inhibitor can be displaced and inhibition can be overcome. It is this property from which the term 'competitive' arises, in that there is competition between substrate and inhibitor for the active site. Malonate is an inhibitor which competes with the substrate succinate for a particular active site; furthermore it cannot react as it cannot be dehydrogenated to the product, fumarate.

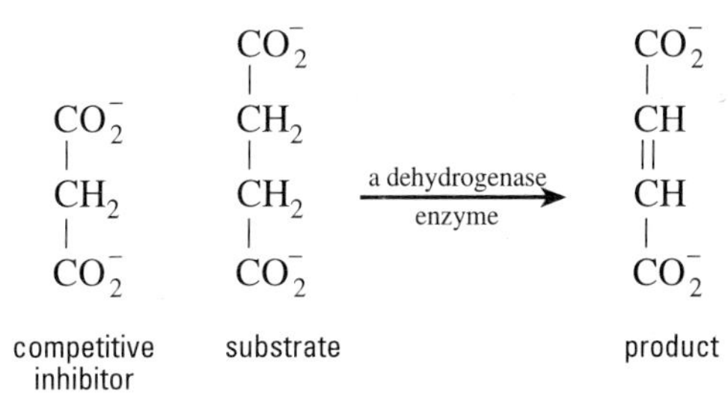

Allosteric inhibitors Finally, some enzymes have separate binding sites for a specific inhibitor. The presence of inhibitor in the inhibitor binding site is believed to alter the three-dimensional shape of the enzyme; the modified enzyme

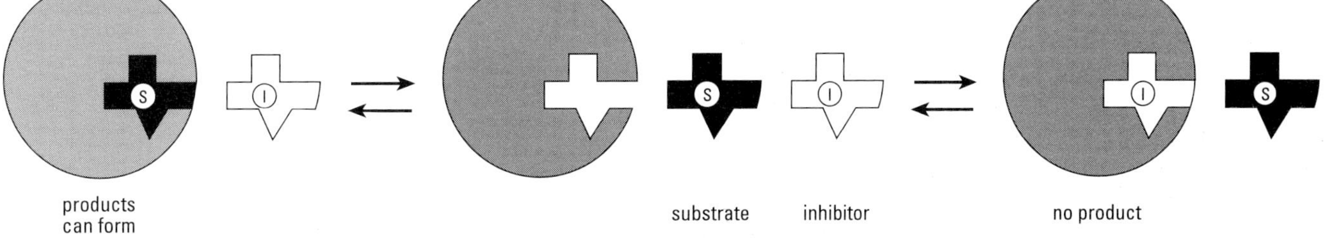

Figure 2.11 Competitive inhibition.

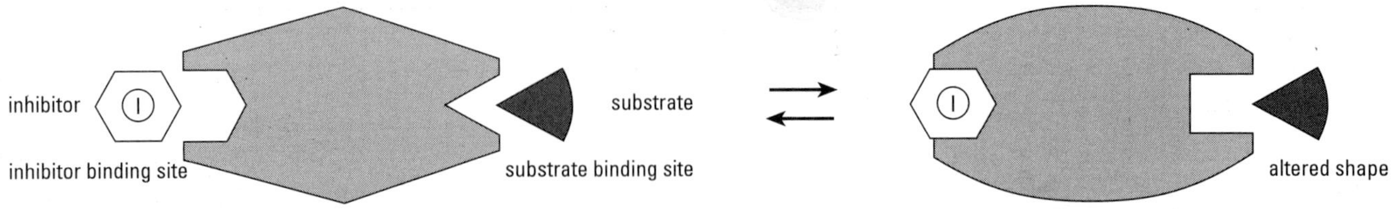

Figure 2.12 Allosteric inhibition. The inhibitor causes a change in shape of the active site and the enzyme becomes inactive.

has reduced catalytic activity. However, since the inhibitor only binds weakly to the binding site, the inhibition is reversible. If the concentration of the inhibitor falls, the enzyme–inhibitor complex dissociates, the optimal three-dimensional shape required for catalysis is re-established, and full enzyme activity is restored. Such inhibition, which plays an important role in the regulation of biochemical processes, is known as **allosteric inhibition** (Greek: *allos stereos* other place).

2.4 Investigating enzymes in action

In this section you will investigate experimentally the properties of two enzymes, amylase and catalase. Factors influencing enzyme activity include pH, the type of buffer solution used, the concentrations of enzyme and substrate, and the temperature. When we wish to study the effect of a particular variable on enzyme activity, all the other factors need to be kept constant.

Experiment 2.4a The effect of pH on enzyme activity

Starch is a mixture of carbohydrate polymers consisting of long chains of glucose units, some with a branched-chain structure. Both types of structure react with iodine to give a characteristic blue–black colour.

Amylase enzyme in our saliva hydrolyses starch producing a complex mixture, which contains mainly maltose. This mixture does not affect the colour of iodine solution although other carbohydrate molecules, produced during the course of the hydrolysis, give rise to violet, brown, and orange colours.

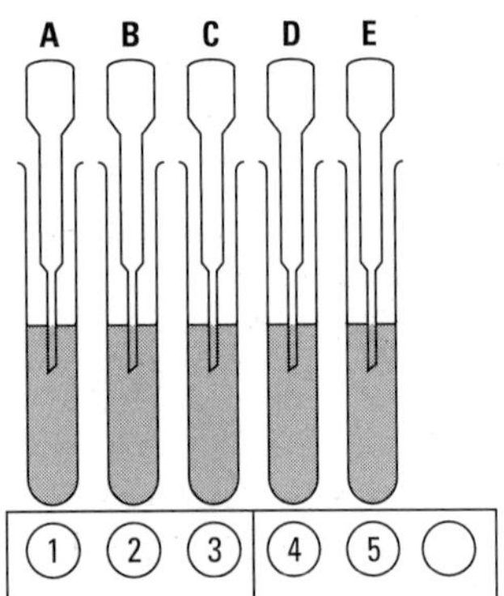

Figure 2.13 Apparatus for experiment 2.4a.

Procedure

⚠ **SAFETY** Saliva can transmit certain infectious diseases. The procedure should be followed carefully and mouth pipettes must not be used. Handle only your own saliva and at the end of the experiment rinse all the apparatus contaminated by your saliva yourself. Then leave the apparatus to soak in freshly-diluted 1% sodium chlorate(I) solution.

Isolation of enzyme

Rinse your mouth out with about 20 cm^3 of pure water into a small beaker. Make sure your saliva and the rinse water are well mixed.

Preliminary measurement to find the approximate level of enzyme activity

1 Put 5 cm^3 of 1% starch, 1 cm^3 of 0.2 M sodium chloride, and 2 cm^3 of 0.067 M phosphate buffer solution, pH 7.0, into a test-tube.
2 Using a dropping pipette, fill each of the hollows in a spotting tile with 5 drops of 0.001 M iodine solution.
3 Using a graduated pipette, fitted with a pipette filler, add 1 cm^3 of the saliva solution to the test-tube. Start the clock, stopper the test-tube and shake it vigorously, and quickly remove a small portion of the mixture, using a dropping pipette. Add 3 drops of the mixture to one of the iodine samples and stir with a glass rod.
At 2 minute intervals, shake the test-tube, withdraw a further 3 drops and add them to a fresh sample of iodine.
Continue until no positive starch test is obtained.
4 Calculate the volume of saliva solution required to hydrolyse most of the starch in about 8 minutes. This volume will be used in subsequent experiments. If necessary concentrate the solution by adding more saliva, or dilute with distilled water.

The effect of pH on amylase activity

1 Set up five test-tubes each containing 5 cm^3 of 1% starch, 1 cm^3 of 0.2 M sodium chloride, and 2 cm^3 of buffer solution of different pH values, as below.

Tube	A	B	C	D	E
pH of buffer solution	4.5	5.9	7.0	8.0	8.8

2 Arrange four spotting tiles as shown and fill each cavity with 5 drops of iodine.
3 Add the calculated volume of saliva solution to test-tube **A**. Start the clock, shake the test-tube, and immediately transfer 3 drops of the mixture to cavity **1**, using a dropping pipette. Stir with a glass rod.
4 At 30 seconds exactly, add a similar volume of saliva solution to test-tube **B**. Shake and immediately transfer 3 drops to cavity **2**, using a clean dropping pipette.
5 At 60, 90, and 120 seconds add saliva to tubes **C**, **D**, and **E**. Test the mixtures by adding 3 drops to cavities **3**, **4**, and **5**.
6 At 150 seconds shake test-tube **A** and transfer a further 3 drops to cavity **6**.
7 At 180 seconds shake test-tube **B** and transfer a further 3 drops to cavity **7**.
8 Remove samples sequentially every 30 seconds until 6 or 7 samples have been removed from each test-tube.
9 Devise a suitable way of recording your results. Remember that the addition of saliva to tubes **A–E** at regular intervals allows the direct comparison of the colours in each row of cavities. For example, the colour in each cavity of row 2 reflects the extent of the reaction in each tube 150 seconds after the addition of enzyme.

Question 1

After the initial iodine-starch colour develops in a cavity, why does it not alter with time?

Question 2

What effect will a change in pH have on the overall charge of a protein molecule? Why do such changes affect enzyme activity?

Experiment 2.4b How does the rate of an enzyme-catalysed reaction vary with the concentration of the substrate?

The enzyme catalase is found in bacterial and animal tissue. In the liver of mammals it catalyses the following reaction:

$$2H_2O_2(aq) \rightarrow 2H_2O(l) + O_2(g)$$

The role of catalase in cells is uncertain but it may prevent the accumulation of poisonous hydrogen peroxide in cells. In this experiment we find out how the rate of a reaction catalysed by an enzyme varies with the concentration of its substrate hydrogen peroxide. The rate of decomposition of hydrogen peroxide is determined by measuring the volume of oxygen produced in a given time.

Procedure

⚠ **SAFETY** 20 volume hydrogen peroxide is irritant. Wear eye protection during this experiment.

Extraction of catalase

1. Weigh out 10 g of liver and cut it into small pieces with scissors. Put the pieces in a mortar and add 3 g of sand. Grind the liver while slowly adding 10 cm^3 of 0.067 M phosphate buffer solution, pH 7. Continue grinding until there are no lumps of liver left.
2. Scrape the homogenate into a depression made in a muslin cloth placed over a 100 cm^3 beaker and extract as much liquid as possible by squeezing.
3. Centrifuge this liquid for 3 minutes. Decant the supernatant into a 10 cm^3 measuring cylinder. Pour half of this supernatant into a 250 cm^3 beaker, make the volume up to 200 cm^3 with phosphate buffer solution, and stir. This is the enzyme extract to be used in experiments **1** to **8** below.
4. Finally take approximately 4 cm^3 of the *diluted* extract, boil it for 1 minute, then cool it to room temperature. Use this boiled extract for experiment **9**.

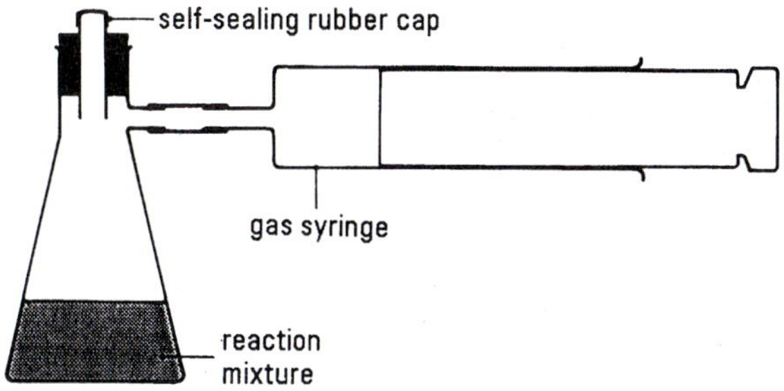

Figure 2.14 Apparatus for experiment 2.4b.

Testing for catalase activity

Set up the apparatus as shown in figure 2.14. The contents of the flasks for each experiment are shown at the bottom of the page. Run the hydrogen peroxide into the flasks, using a burette.

1. The same flask may be used for the nine separate experiments, provided it is thoroughly washed with distilled water and excess drops of water are removed by shaking.
2. Disconnect the self-sealing cap and push the plunger of the gas syringe to zero before starting the experiment. Replace the self-sealing cap. Fit a hypodermic needle to a 2 cm^3 syringe. Draw 2 cm^3 of the enzyme solution into the hypodermic syringe, making sure that all air is expelled from the syringe.
3. Start the reaction by piercing the self-sealing cap with the hypodermic needle and *rapidly* injecting the enzyme solution into the flask. After all the enzyme has been injected, *immediately* remove the needle, start the stop clock, and shake the flask vigorously.
4. Measure the volume of oxygen produced after 15 seconds, keeping the flask vigorously swirled all the time. If the gas-syringe plunger appears to stick, ease the plunger out slightly by hand. It may be necessary to adjust the concentration of enzyme in experiment **1** so that a suitable volume of oxygen is produced in 15 seconds. This volume should be between 40 cm^3 and 60 cm^3.
5. Repeat this procedure for experiments **2** to **8**, washing the flasks thoroughly with distilled water between experiments and shaking well to get rid of the water. Lastly, using a fresh hypodermic needle and syringe, repeat the procedure (experiment **9**), but in this case inject 2 cm^3 of the boiled enzyme solution.

Treatment of results

1. Subtract a volume of 2 cm^3 from each reading of the gas syringe. After this correction has been made, subtract the readings of the control experiment (**8**) from the gas syringe readings.
2. Calculate the initial molarity of hydrogen peroxide in each experiment. '20 volume' H_2O_2 has a concentration of 1.74 moles per dm^3, assuming that no decomposition has occurred.
3. Plot a graph of volume of oxygen released (proportional to the rate of reaction) against the concentration of substrate (hydrogen peroxide).

Question 3

Why was a volume of 2 cm^3 subtracted from each reading of the gas syringe?

Question 4

What was the purpose of experiment **8**?

Question 5

What was the effect of boiling the catalase? Using your knowledge of protein structure, give an explanation of this effect.

Question 6

How does the rate of reaction depend on substrate concentration at

i relatively low substrate concentrations,
ii high substrate concentrations?

Question 7

What prevents the enzyme from working at its highest catalytic activity at low substrate concentrations?

Question 8

At high substrate concentrations, why does a further increase in substrate concentration have little effect on the rate of the enzyme-catalysed reaction?

Experiment	1	2	3	4	5	6	7	8	9
Volume of 0.067 M phosphate buffer solution, pH 7 /cm^3	10	10	10	10	10	10	10	10	10
Volume of '20 volume' H_2O_2/cm^3	15	10	7	5	3.5	2	1	0	15
Volume of distilled water /cm^3	3	8	11	13	14.5	16	17	18	3

2.5 Immobilized enzymes in industry

Although the production of products like bread and beer predates our knowledge of even the existence of enzymes in the processes involved, the modern use of enzymes has revolutionized many industrial processes and the quality of their products. The benefits brought to industry by the use of enzymes are increased productivity, lowered costs, and reduced amounts of chemical waste for disposal.

Batch reactions using free, soluble enzymes are used in many industrial processes. The enzyme and the substrate are mixed at a suitable temperature in a reactor vessel and left until the required amount of product has been formed. Generally no attempt is made to recover the enzyme because it is too costly to do so. In many cases the enzyme preparations are expensive, and so only small quantities are used in each batch. Lengthy reaction times are therefore necessary and this may result in the extensive formation of unwanted products due to side reactions.

For prolonged use in a continuous process or repeated use in a batch process, an enzyme needs to be immobilized, that is, attached to a solid surface so that it can be separated from the reaction mixture. Initially the cost of producing an immobilized enzyme system is greater than that of producing a soluble enzyme preparation for the same reaction. However, the insoluble enzyme may be packed into columns and used continuously for long periods. Alternatively, the enzyme may be re-isolated simply by centrifugation or filtration after use in a batch reactor; it may then be re-used several times. Higher enzyme concentrations are possible and shorter reaction times reduce both the operational costs and the need for extensive purification of the product.

Methods of immobilization

Many methods of immobilizing soluble enzymes have been attempted, with varying degrees of success. The enzyme can be bound to a solid support, either by weak intermolecular forces or by strong covalent bonding.

Adsorption onto an insoluble support, such as porous glass

The protein is held in place by ionic interactions between charged amino acid side chains and charged groups on the support material, and van der Waals interactions between non-polar amino acid side chains and non-polar areas of the support. The immobilized enzyme is easily produced but since the bonding between the enzyme and the support system is weak, there is a danger that the enzyme may be washed off the support.

Trapping the enzyme within a gel such as collagen

A solution containing the enzyme and a water-soluble monomer is prepared and the monomer is allowed to polymerize. The large enzyme molecules are trapped within a complex network of polymer molecules, but small molecules such as the substrate are able to penetrate the network to interact with the enzyme. Therefore enzymes which act on substrates of high molar mass cannot be used. This technique is inexpensive, and the immobilized enzymes can often be used for a long time.

Covalent binding onto a solid support such as cellulose or nylon

One or more of the enzyme side chains, containing usually an $—NH_2$, —OH, or $—CO_2H$ group, react covalently with the support material. The enzyme can show increased stability, and its active site remains readily accessible to the substrate. However, the production of a suitable solid support with reactive groups can be expensive.

Encapsulation behind a permeable membrane, for example a membrane of nylon

The enzyme is enclosed in a membrane which allows substrate and product molecules to pass freely but will retain large enzyme molecules. Production of the immobilized enzyme can be expensive but this is a good method of recovering product free from enzyme.

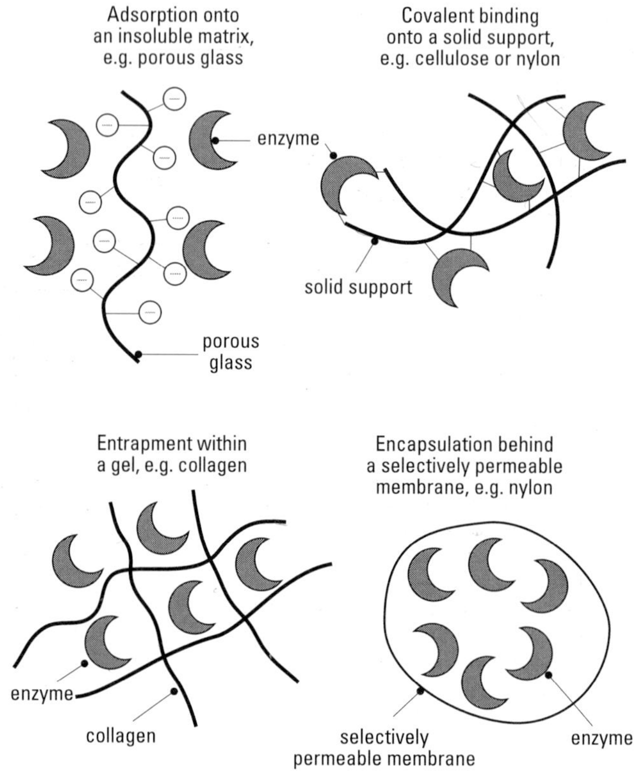

Figure 2.15 Methods of enzyme immobilization.

The separation of the chiral forms of an amino acid

When an amino acid is prepared by chemical synthesis the result is usually a mixture of both chiral forms, the D-form and the L-form. An enzyme can be used to isolate the biologically useful L-amino acid from a mixture of the acyl D- and L-amino acids.

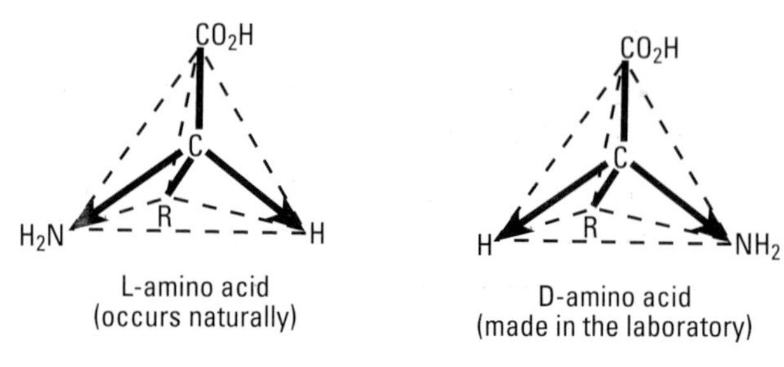

The enzyme used in the process is extracted from the fungus *Aspergillus*. The enzyme is adsorbed onto a chemically treated polysaccharide so that it is held in place by ionic bonds. The immobilized enzyme is packed in columns and can be used continuously for several weeks. The specificity of the enzyme is unaffected by immobilization. The optimum temperature for enzyme activity is increased from 60 °C to 72 °C but the activity is reduced by about half. This loss of activity, however, is more than compensated by the fact that the enzyme can be used for extended periods. Indeed L-amino acids can be produced by the continuous method at about 60% of the cost of the batch process, using free soluble enzyme.

The production of high fructose syrup

Sucrose, ordinary sugar, is our most commonly used sweetener but for use in the food industry starch is a cheaper source than sugar cane. However, the hydrolysis of starch by acids produces glucose, which has only about 40% of the sweetness of sucrose. But glucose can be converted to fructose which is considerably sweeter than sucrose.

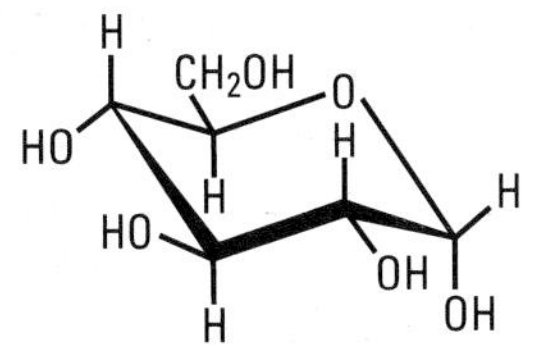

glucose

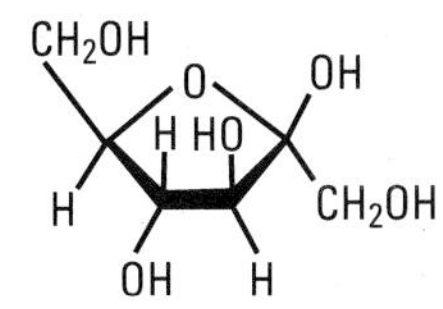

fructose

In the production of high fructose syrup, starch from corn is hydrolysed to glucose, which is then partially converted to fructose using an enzyme. The result is a mixture of glucose and fructose which is similar in sweetness to sugar.

Enzymes which convert glucose to fructose are readily available from bacteria and the enzyme glucose isomerase has proved easy to immobilize. The immobilized enzyme is used in continuous reactors at 60–65 °C and pH 7 for over 1000 hours. The process has dramatically reduced the cost of producing high fructose syrup.

Experiment 2.5 Making whey syrup using an immobilized enzyme

Whey is a by-product from cheese making, but the sugar it contains, lactose, is not very sweet. The hydrolysis of lactose, using enzyme, produces glucose and galactose, which are much sweeter and make the product, whey syrup, useful in the confectionary trade.

Procedure

1. Add 2 cm^3 of lactase enzyme solution to 8 cm^3 of a 2% sodium alginate solution, mix well and draw up into a plastic syringe (without the needle).
2. Allow drops of the mixture fall into 100 cm^3 of 0.1 M calcium chloride solution so that small beads are formed. Let the beads set for a few minutes. Collect the beads in a small strainer and wash free of calcium chloride solution.
3. Put a circle of nylon gauze at the bottom of a 10 cm^3 syringe fitted with a stop-tap to prevent the beads blocking the nozzle (see figure 2.16). Suspend the beads in water and pour into the syringe.
4. Pass 50 cm^3 of an 8% whey solution through the column of enzyme beads and test the effluent at regular intervals with semi-quantitative diabetic glucose test strips. Finish by washing the column free of whey solution.
5. Plot a graph of your results.

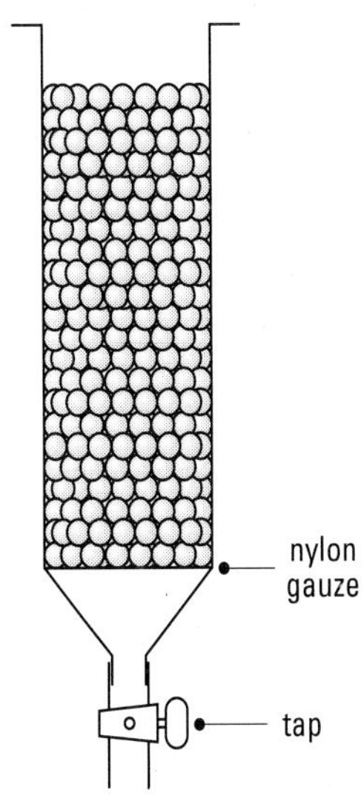

Figure 2.16 Apparatus for experiment 2.5.

Question 9

What can you deduce from your graph? Make an estimate of the concentration of glucose in your whey syrup.

Question 10

Can the column be used again?

Question 11

What type of immobilization was involved in this experiment?

REVIEW TASK

When you have finished reading this chapter and completed the practical work you should make your own notes. Try to include an example to illustrate each point and make sketches or draw up tables when you think that will be helpful to you in remembering and understanding the material.

In writing your notes pay particular attention to the following points:

- what are the main components of a cell and their biochemical functions?
- how are enzymes obtained from cells with the minimum disruption to their properties?
- what is the primary, secondary and tertiary structure of a protein and how do the bonds involved help to maintain the structure?
- how does an enzyme work and how does your model explain the specificity of enzymes and the effects which inhibitors, changes in pH, in substrate concentration and of temperature have on enzyme activity?
- why and how are enzymes immobilized?
- lastly, you should be able to describe the experiments in this chapter and be able to explain the procedures used.

CHAPTER 3 METABOLISM

The chemical processes which take place in living organisms are given the general term **metabolism**. Many of these processes are enzyme-catalysed reactions which involve the hydrolysis and oxidation of the compounds in food to simpler molecules. Other processes are concerned with the production of complex molecules from the simple molecules obtained from food. Living organisms have to achieve a difficult energy balance with the digestion of food to simpler molecules likely to have a favourable total entropy change (ΔS_{total} positive) while the building of complex molecules will have an unfavourable entropy change (ΔS_{total} negative). If you consider the comment by Clausius

'The entropy of the universe tends to increase'

you will appreciate the entropy challenge that the evolution of life had to solve.

The movement of molecules through the reaction steps in living organisms is summarized in **metabolic pathways**. Each metabolic pathway consists of a sequence of chemical reactions, and each reaction is catalysed by a specific enzyme. There are four important aspects:

- digestion and absorption: food is broken down into simpler molecules which are small enough to pass into the bloodstream
- synthesis of biochemical compounds: some of the simpler molecules produced by digestion are converted to other compounds, such as acetyl-CoA which plays a central role in metabolism
- reactions which are energetically favourable: these reactions provide the energy required for other metabolic processes: the overall result is often the oxidation of organic compounds to carbon dioxide and water
- reactions which are energetically unfavourable: these reactions are usually involved in the synthesis of compounds with complex structures, such as proteins.

3.1 The oxidation of food

The main components of the diet of mammals such as ourselves are carbohydrate, protein and fat. The carbohydrates in our diet are the main source of energy for all our living cells (and the same is true for all plants and animals). The carbohydrate glucose can be oxidized in almost all cells, in reactions involving the use of enzymes as catalysts. When glucose is fully oxidized, a large amount of energy is liberated (data at 298 K).

$$C_6H_{12}O_6(s) + 6O_2(g) \longrightarrow 6CO_2(g) + 6H_2O(l)$$
$$\Delta H = -2800\ \text{kJ mol}^{-1}$$
$$\Delta S_{total} = +9600\ \text{J mol}^{-1}\text{K}^{-1}$$

However, the energy made available by the catalytic oxidation of glucose in cells is not all liberated as heat energy, although some heat energy is necessary to maintain the body temperature of mammals and birds. Some of the available energy is needed for doing mechanical work, for instance, in muscle cells. During growth and development, much energy is required for the synthesis of cell constituents. In mature cells the chemical composition remains relatively constant, but there is still a constant turnover of many components with energy needed for their synthesis.

This, and the next, section considers how the energy from the oxidation of food is made available for synthetic reactions.

The oxidation of food can be divided into three main stages:

1. the hydrolysis of food into smaller molecules, such as glucose
2. the further breakdown of these molecules within individual cells
3. the oxidation of compounds in the **citric acid cycle** to carbon dioxide.

These three stages will be considered in turn.

1 The hydrolysis of food

The main bulk of food is in polymer form as polysaccharides (carbohydrate), protein, or triglycerides (fats). These large molecules must be split into smaller molecules by enzymes in the stomach and small intestine before they can be absorbed. Proteins are split to peptides and eventually to amino acids by proteases. Polysaccharides are hydrolysed with the help of the enzyme amylase to disaccharides, and thence to monosaccharides (mainly glucose). Fats are digested by the enzyme lipase, which produces fatty acids (carboxylic acids) and glycerol (propane-1,2,3-triol). Most of these products are soluble in water whereas many of the original food polymers were insoluble.

2 The production of acetyl-CoA from glucose

First glucose is converted to pyruvic acid. This whole process, the conversion of glucose to pyruvic acid, is known as **glycolysis**.

There are ten steps in the pathway, each of which is catalysed by a specific enzyme, but only one of the steps is an oxidation reaction.

$$\underset{(C_6H_{12}O_6)}{\text{glucose}} \xrightarrow{\text{10 steps}} \underset{(2C_3H_3O_3^-)}{\text{2 pyruvate}}$$

In cells at a pH of about 7.2 pyruvic acid exists as the pyruvate anion and is called 'pyruvate'.

Next the pyruvate formed from glucose reacts to form **acetyl-CoA**. Acetyl-CoA is the name given to a compound formed when a complex organic molecule, coenzyme A, combines with an ethanoyl group, CH_3CO—, sometimes called the **acetyl group**. The formation of acetyl-CoA from pyruvate is an oxidation reaction. Carbon dioxide is also produced.

So from each mole of glucose (remember 1 mole of glucose produced 2 moles of pyruvate):

$$\underset{(2C_3H_3O_3^-)}{2\ \text{pyruvate}} + \underset{(2CoA)}{2\ \text{coenzyme-A}} \longrightarrow \underset{(2CH_3CO\text{-}CoA)}{2\ \text{acetyl-CoA}} + \underset{(2CO_2)}{2\ \text{carbon dioxide}}$$

Comment

Some living organisms such as yeasts are able to survive when grown in very limited supplies of oxygen (anaerobic conditions). In yeast cells the pyruvate formed during glycolysis is decarboxylated, forming ethanal, which is then reduced to ethanol. The overall process is the fermentation of glucose to alcohol:

glucose → 2 pyruvate → 2 ethanal + 2 carbon dioxide → 2 ethanol

During vigorous exercise we may not have enough oxygen supplied to our muscles (anaerobic conditions), and pyruvate can be converted to lactate instead of entering the citric acid cycle. Lactate in our muscles produces the characteristic symptoms of cramp.

$$\underset{\text{pyruvate}}{CH_3COCO_2^-} \longrightarrow \underset{\text{lactate}}{CH_3CHOHCO_2^-}$$

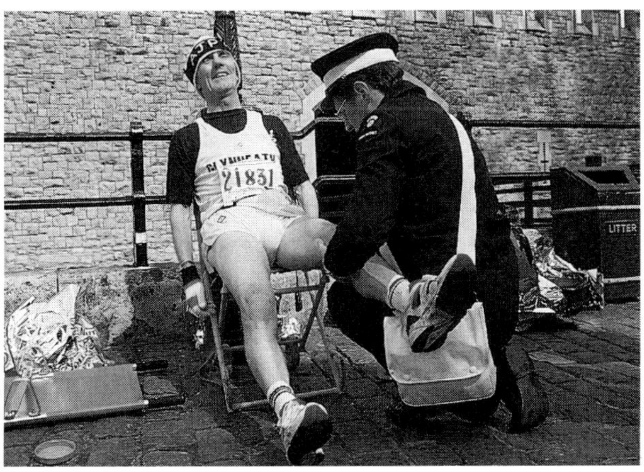

Figure 3.1 We can get cramp when our muscle cells have insufficient oxygen.

When the normal supply of oxygen is re-established, the accumulated lactate is oxidized back to pyruvate.

REVIEW TASK

Revise your main course organic chemistry by planning synthetic routes from

1 ethanal to ethanol
2 lactic acid to pyruvate.

Use your textbook as a source of information.

3 The oxidation of acetyl-CoA in the citric acid cycle

Acetyl-CoA can be used as the building block for many useful biochemicals, particularly fatty acids, but most acetyl-CoA enters the citric acid cycle. The remaining carbon from the glucose is now oxidized to carbon dioxide and the coenzyme A is regenerated for reuse in a cycle involving eight steps. From each molecule of glucose:

$$2CH_3CO\text{-}CoA \xrightarrow{\text{8 steps}} 4CO_2 + 2CoA$$

The overall result of the glycolysis pathway and the citric acid cycle is the complete oxidation of glucose to carbon dioxide and water (see figure 3.2).

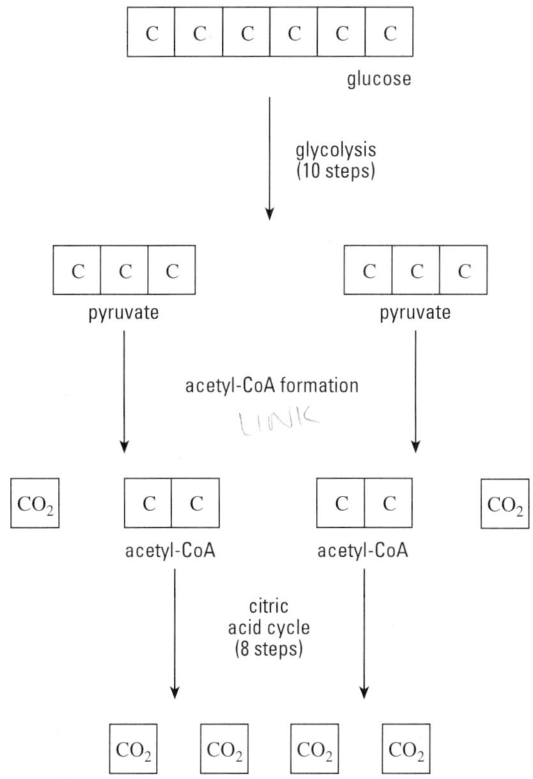

Figure 3.2 The complete oxidation of glucose via the glycolysis pathway and the citric acid cycle.

The role of NAD in metabolic oxidation reactions

The oxidation reactions which occur during the breakdown of glucose from foods are mostly catalysed by enzymes known as **dehydrogenases**. These are enzymes which catalyse the removal of two hydrogen atoms from a substrate molecule by an oxidizing agent.

Although oxygen from the air is the ultimate source for metabolic oxidation reactions, the actual oxidizing agent involved in most of these reactions is a compound called **NAD** (**n**icotinamide **a**denine **d**inucleotide). NAD is the general name of the compound, with NAD^+ used for the oxidized form and NADH for the reduced form of the molecule. NAD^+ was the oxidizing agent in the oxidation of pyruvate to acetyl-CoA and carbon dioxide that was mentioned above.

$$C_3H_3O_3^- + CoA + NAD^+ \longrightarrow CH_3CO\text{-}CoA + CO_2 + NADH$$

The complete oxidation of 1 mole of glucose uses 10 moles of NAD^+ (and 2 moles of another oxidizing agent). So the products will include 10 moles of the reducing agent NADH as well as carbon dioxide and some heat energy. This is an important advantage of the metabolic pathway compared to simple combustion, which would have produced carbon dioxide, water, 2800 kilojoules and nothing else.

What is the advantage in having 10 moles of NADH available? We will answer this question in the next section.

3.2 Energy and the role of ATP

In all living organisms a particular compound, **ATP** (**a**denosine **tri**phosphate), has a critical role to play in making energy available for metabolic reactions. ATP is a **nucleotide** and its structure is shown in figure 3.3. It consists of three **phosphate groups** linked in a short chain (polyphosphate ions are quite a common inorganic structure) and then covalently bonded to the hydroxyl group of a sugar, ribose. The last part of the molecule is adenine, an organic base.

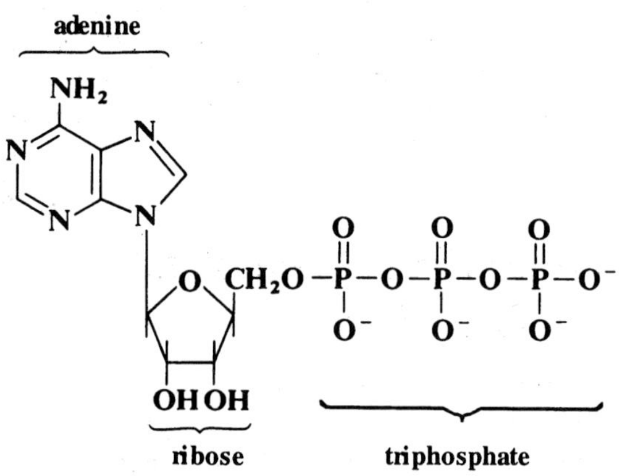

Figure 3.3 The structure of ATP, adenosine triphosphate.

ATP is hydrolysed to **ADP** (**a**denosine **di**phosphate) and an inorganic phosphate ion in a reaction that is energetically favourable. Energy is required to break bonds between phosphate groups, and in water, but there is a net gain of energy when the products form (see below).

ATP and ADP are anions but the negative charges are not normally shown, so their equations appear to be unbalanced. The equation can also be simplified by writing Ⓟ for PO_3 groups.

In many chemical transformations and synthetic reactions involving enzyme catalysts ATP plays an essential role. Metabolic pathways are using large amounts of ATP all the time so its synthesis is an important process. Ultimately each ATP dependent process produces ADP and a key step in metabolism is the regeneration of ATP. This reaction is of course energetically unfavourable and is where the energy available from the oxidation of food is needed.

ATP synthesis

The synthesis of ATP involves the reaction

$$\text{adenosine–}\textcircled{P}^{-}\text{—}\textcircled{P}^{2-} + H^+ + HPO_4^{2-} \longrightarrow \text{adenosine–}\textcircled{P}^{-}\text{—}\textcircled{P}^{-}\text{—}\textcircled{P}^{2-} + H_2O$$

$$\Delta S_{total} = -107\ \text{J mol}^{-1}\text{K}^{-1}$$

This process is called **phosphorylation** because it involves adding a phosphate group. Under standard conditions the reaction is energetically unfavourable, but in the cell it is linked to the oxidation of NADH back to NAD^+. The two processes are said to be **coupled** in a process called **oxidative phosphorylation** and are illustrated in figure 3.4.

A second way in which phosphorylation is achieved is called **substrate level phosphorylation**. In glycolysis and the citric acid cycle, several enzyme-catalysed reactions lead to the phosphorylation of ADP. In the glycolysis pathway one of the intermediate compounds formed is phospho-enolpyruvate. This compound reacts with ADP, in the presence of an enzyme, to form ATP and pyruvate.

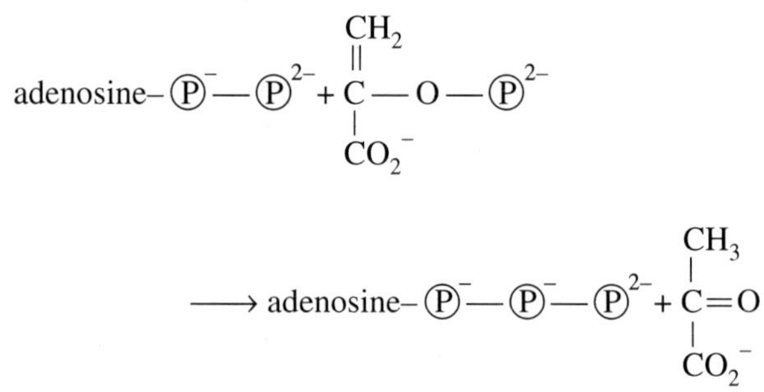

ADP + phospho-enolpyruvate ⟶ ATP + pyruvate

ATP production from the oxidation of glucose

The main function of glycolysis and the citric acid cycle is therefore to produce ATP for use in energetically unfavourable processes. The complete oxidation of 1 mole of glucose

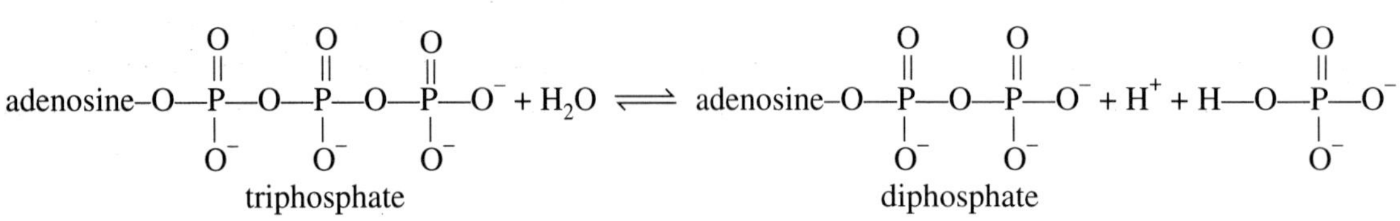

$$ATP + H_2O \rightleftharpoons ADP + H^+ + HPO_4^{2-} \qquad \Delta S_{total} = +107\ \text{J mol}^{-1}\text{K}^{-1}$$

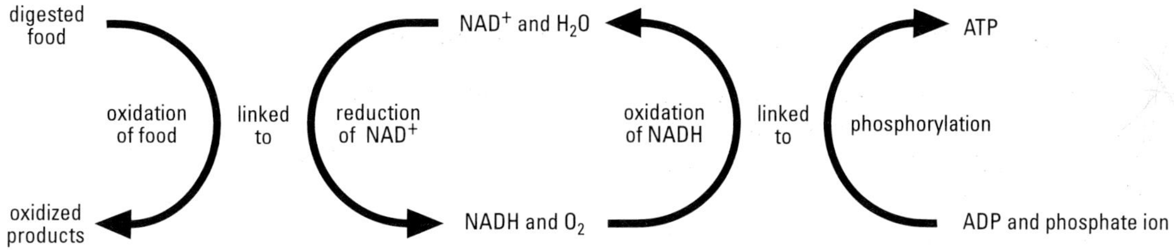

Figure 3.4 The synthesis of ATP and NAD^+ in coupled reactions, by the process called oxidative phosphorylation.

produces 10 moles of NADH which can be converted to 30 moles of ATP through their coupled reactions. In addition, other processes yield a further 8 moles of ATP. Thus the maximum amount of ATP that can be produced from the oxidation of 1 mole of glucose is 30 + 8 = 38 moles.

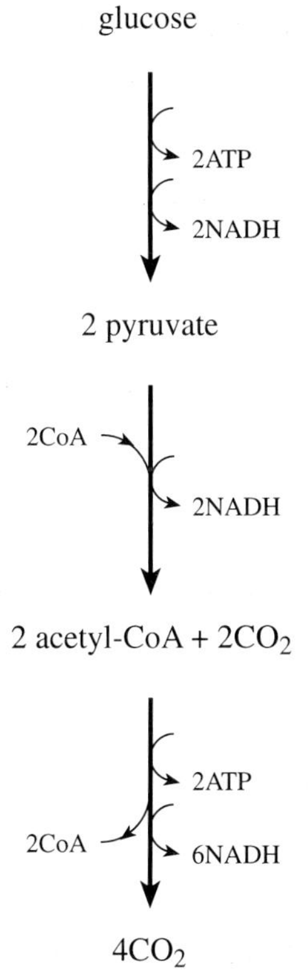

Figure 3.5 The involvement of NADH/ATP in the oxidation of glucose.

Plants also make ATP from ADP and phosphate ions but the energy can also come from sunlight; whereas in animals the energy comes solely from the oxidation of food.

If cellular activities are to be controlled, spontaneous reactions must not occur in the absence of enzymes. The intermediates formed during metabolism must be kinetically inert under physiological conditions. For instance, the hydrolysis of ATP to ADP and phosphate is energetically favourable (ΔS_{total} = +107 J $mol^{-1}K^{-1}$) but the rate of reaction is very slow because of the high activation energy of the reaction, so spontaneous hydrolysis does not occur. This ensures that the ATP produced during the oxidation of food is available for cellular processes. The role of ATP in metabolism is summarized in figure 3.6.

3.3 Metabolic pathways and metabolic control

The links between some metabolic pathways for the main components of our food are illustrated in figure 3.7, page 19.

The protein in our food is hydrolysed to amino acids. Some amino acids can be synthesized in metabolic processes, but others cannot and we can only obtain them from our food. Amino acids which must be supplied in our food are called **essential amino acids**.

Amino acids are needed for the synthesis of proteins, for example, enzymes to catalyse reactions and structural proteins such as hair and nails. They are also used to make the nitrogenous organic bases which are needed for the synthesis of DNA and RNA. Amino acids, once the amine group has been removed, can enter metabolic pathways at a number of different points: for the sake of simplicity only two examples, leucine and alanine, are shown.

Fats, after hydrolysis to fatty acids and glycerol, can be oxidized to acetyl-CoA, a process which releases considerable amounts of energy. The acetyl-CoA can be used for synthetic reactions or further oxidized in the citric acid cycle to produce more energy. Fats release more energy per gram than either proteins or carbohydrates; if fatty acids and glycerol are not needed when they are first absorbed, they can be stored as an energy reserve by reforming fats.

Notice that acetyl-CoA can be formed in three ways: from glucose, from fatty acids, and from some amino acids, following the loss of their amine groups.

You should also notice that many reactions and processes are reversible but some are not. This has important implications. For instance pyruvate cannot be made from acetyl-CoA and

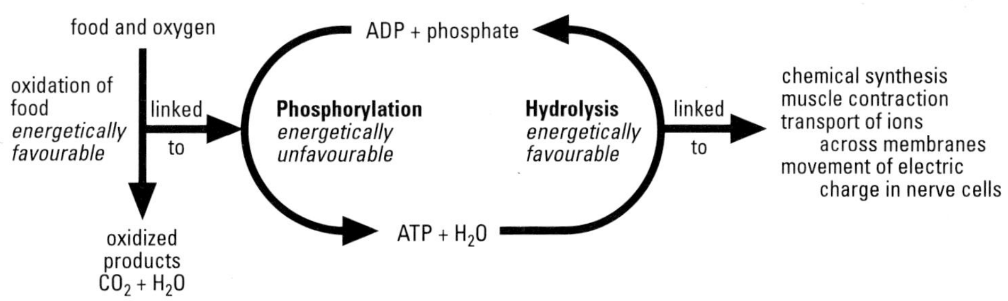

Figure 3.6 The role of ATP in metabolism.

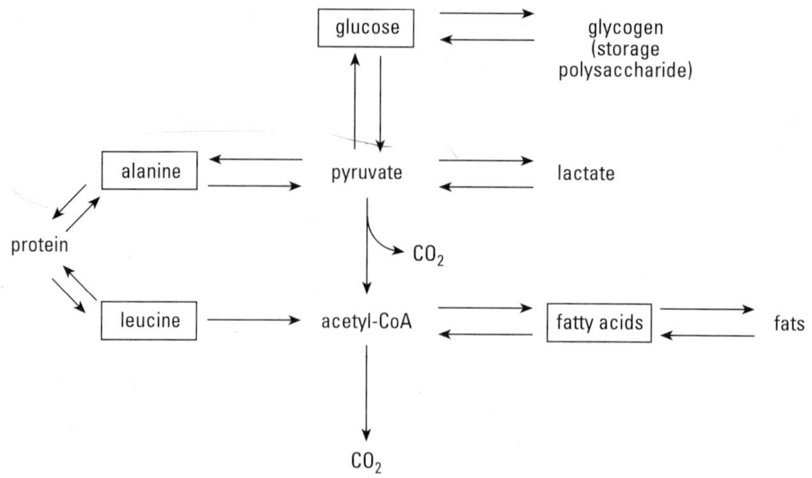

Figure 3.7 Links in metabolic pathways: the metabolism of glucose, fatty acids and two amino acids.

this means that carbohydrates cannot be produced from fatty acids or from some amino acids, such as leucine. Similarly leucine cannot be made from acetyl-CoA and for healthy growth must be available from our food. In contrast other amino acids, like alanine, can be synthesized from glucose and are therefore not classified as essential amino acids.

The interconversion of sugars, amino acids, and fats through the links between metabolic pathways is of considerable importance to cells with very specific requirements, particularly when the variety of food is limited. The human brain, for example, needs 120 g of glucose per day. The total readily available in the body fluids and in glycogen, the glucose storage polymer, does not amount to much more than 200 g. Thus during periods of starvation or low carbohydrate intake the production of glucose from non-carbohydrate sources is essential.

MEDICAL CASE STUDY Phenylketonuria

Phenylketonuria is a disorder of amino acid metabolism. It results from the reduced activity of the enzyme which converts phenylalanine to tyrosine in the liver. It is an inherited condition and 1 in 7000 children in the UK are affected. The condition causes severe mental retardation and the children may have a characteristic mousy odour due to the presence of phenylethanoic acid.

Even though the condition is quite rare prompt treatment at birth can prevent the damage to the nervous system. This justifies routine screening of all babies at the end of the first week of life for phenylalanine in their blood.

Babies with the condition must be put on a diet low in phenylalanine as soon as possible to protect the brain. However, sufficient phenylalanine must be provided for growth as it is an essential amino acid. Unfortunately the diet tastes awful.

After the age of eight the diet is less important as the nervous system is less vulnerable. If a woman with phenylketonuria is pregnant even tighter dietary control is necessary if the fetus is not to suffer.

Metabolic control

The DNA within the simplest cell contains enough information to direct the synthesis of many enzymes, each of which can catalyse a reaction or, exceptionally, more than one. Some of these enzymes are involved in the oxidation of food. Others catalyse the synthesis of the thousands of different proteins, carbohydrates, fats, and smaller organic molecules needed for the healthy survival of an organism. The exact requirements of a particular cell, both for energy and for the products of synthetic reactions, will be continually changing. Therefore, the flow of material through the intricate network of chemical reactions must be carefully controlled if the individual needs of a cell are to be supplied at all times.

Metabolic control may be exerted in several ways, but one possible method involves the control of enzyme activities within a cell.

The control of a biosynthetic pathway

As our example of metabolic control through enzyme activity we shall consider the synthesis of haem, the structure of which is shown in figure 3.8. Haem molecules are normally bound to specific protein molecules forming haemproteins. Haemproteins have an important role to play in many life processes. For example, haemoglobin is involved in oxygen transport, so a sufficient supply of haem is essential for the survival of an organism.

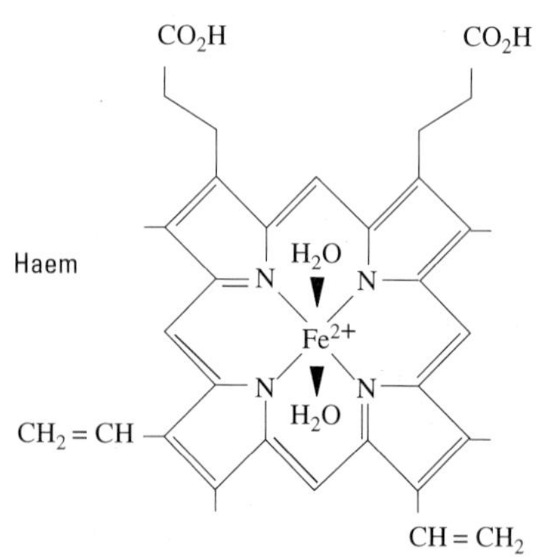

Figure 3.8a The structure of haem. Haem is an example of a transition metal complex involving iron(II).

Figure 3.8b The structure of haemoglobin.

MEDICAL CASE STUDY Poisoning by cyanide

Hydrogen cyanide gas and cyanide ions in solution are very dangerous and rapid acting poisons. Some fruit stones contain compounds that are converted to cyanide during digestion and can cause accidental cyanide poisoning. Headache and nausea are followed by convulsions and coma. Death may occur in hours but treatment is generally effective if an antidote is given quickly.

Cyanide forms stable complexes with iron(III) ions and will rapidly bond to the iron(III) of a haem group in a vital enzyme (cytochrome). Metabolic reactions that involve oxidation cease and cells soon stop functioning.

Administration of oxygen is the only first aid treatment that is safe. In hospital possible treatments include the injection of nitrites, which convert oxyhaemoglobin containing iron(II), to a form of haemoglobin containing iron(III). The iron(III) form of haemoglobin forms a more stable cyanide complex and the cyanide is therefore not able to disrupt essential metabolic reactions. Patients can also be treated with thiosulphate, $S_2O_3^{2-}$, which will react with cyanide to form thiocyanate, CNS^-, which is relatively non-toxic and readily excreted.

The cells of nearly all organisms are found to contain appreciable quantities of several haemproteins, although the concentration of free haem is generally maintained at a very low level. The net synthesis of haem needed in any tissue is extremely small; it has been estimated that the total hourly requirement for haem synthesis per gram of rat liver is less than 2×10^{-8} moles. Lower production of haem has serious consequences, reducing the ability of the animal to withstand infection. Conversely, overproduction may be equally undesirable; high concentrations of unwanted haem disrupts cell function seriously, and the excretion of the unwanted haem from the liver can affect tissues all over the body.

In the liver haem is synthesized in six enzymic steps from two readily available substrates. These are succinyl-CoA, an intermediate in the citric acid cycle, and glycine, an amino acid available from the hydrolysis of protein. The production of the first intermediate, **ALA** (5-**a**mino **l**aevulinic **a**cid), is catalysed by the enzyme **ALA synthetase**. This enzyme has a number of interesting properties.

1 The activity of the enzyme is much lower than any of the subsequent enzymes in the pathway. Thus this first step in the pathway is the rate determining step.
2 The enzyme is inhibited by haem. This is an example of allosteric inhibition (see section 2.3).
3 The enzyme is very unstable. It has been shown that the half-life of ALA synthetase is about 70 minutes. The other enzymes in the pathway have half-lives of several days.
4 The rate of synthesis of the enzyme from amino acids is reduced by the presence of free haem, an example of 'self-inhibition'.
5 The rate of synthesis of the enzyme in the liver is increased by a variety of drugs (particularly barbiturates).

The production of ALA seems to be the step that controls the rate of production of haem. Any ALA produced is rapidly converted to haem with no accumulation of intermediates.

Figure 3.9 The synthesis of haem.

There would seem to be two possible mechanisms by which the pathway might be controlled: either by modifying the activity of the ALA synthetase molecules or by altering their concentration. The experimental evidence suggests that both possibilities occur in mammalian tissue. Haem, the end product of this particular pathway, both inhibits ALA synthetase activity and represses the formation of new enzyme. In this way it directly controls the rate of its own synthesis.

The short half-life of the enzyme ensures a rapid response to new situations. ALA synthetase levels can rise rapidly in the presence of drugs. Conversely, accumulation of haem can lead to a significant reduction in the concentration of the enzyme within hours.

STUDY TASK Haem synthesis

Copy figure 3.9 into your notes and use the description of the properties of the enzyme ALA synthetase to develop the diagram to illustrate the control of the synthesis of haem by:

1 the synthesis of the enzyme from amino acids
2 the degradation of the enzyme to amino acids
3 the effect of haem on the enzyme's synthesis
4 the inhibition of the enzyme by haem
5 the presence of barbiturates.

The action of hormones

A second level of control is needed in higher organisms because the activities of different cells must be co-ordinated. This second level of control is exerted through the action of **hormones**. Although many hormones do not enter cells but interact with the cell membrane, they can affect the concentration or activity of enzymes within 'target' cells. Hormones can alter the behaviour of a cell wall, the rate of synthesis of a particular enzyme, or the activity of an enzyme.

Insulin is an example of a hormone; it is released from the pancreas when blood sugar levels are high. The presence of insulin in the bloodstream increases the passage of glucose through certain cell membranes. In muscles it stimulates glycogen formation, while in the liver it stimulates glycolysis, fatty acid synthesis, and glycogen formation. Also the synthesis in the liver of glucose from pyruvate is reduced. All these processes (see figure 3.7) accelerate the metabolism of glucose and encourage the removal of sugar from the blood. Diabetes is often caused by a failure of the pancreas to produce sufficient active insulin.

Control of metabolism

The following general principles seem to apply:

- the rate-controlling step is generally the first step of a pathway
- the activity of the enzyme catalysing this step is increased or decreased by a variety of activators or inhibitors: often inhibition is brought about by the end product
- in some cases, the enzyme catalysing the rate-determining step has a short half-life, so the concentration of the enzyme may be affected by a change in the rate of production
- hormones co-ordinate the activities of different cells.

Experiment 3.3 Investigating the synthesis of starch by a metabolic reaction

You are going to use the enzyme **starch phosphorylase** to try to prepare starch from glucose and some closely related compounds. Starch is easily detected by its reaction with aqueous iodine to give a blue–black colour.

The enzyme used in this experiment is obtained form potato cells. However, potatoes already contain starch so the enzyme extract will have to be tested to check that it contains no starch.

CARE The glucose-1-phosphate solution you are using is quickly hydrolysed. Keep the solution in an ice bath.

Procedure

1. Place a 2 cm cube of potato in a mortar with 10 cm^3 of pure water and cut into small pieces. Add a little *clean* fine sand and grind to a fine pulp. Filter the pulp through a fluted filter paper of qualitative grade that has been moistened with water; discard the first two drops of filtrate.
2. Test the filtrate for the absence of starch by mixing 1 drop with 1 drop of 0.005 M iodine solution. If the mixture turns blue filter the extract a second time.
3. Put 1 drop of a 0.05 M solution of glucose in each cavity (five is sufficient) of a row on a spotting tile; repeat using sucrose and glucose-1-phosphate solutions in two other rows.
4. Add 1 drop of filtered and starch-free potato extract to each cavity on your spotting tile.
5. At 5 minute intervals add iodine solution to one cavity in each row. Record any colour changes.
6. Carry out a blank test (a 'control') by adding iodine solution to drops of the solutions of glucose, sucrose and glucose-1-phosphate. At the end of the experiment, repeat the test on the potato extract.

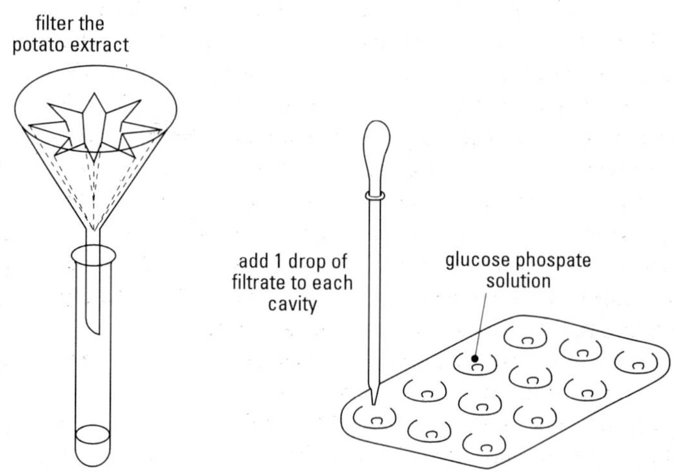

Figure 3.10

Question 1

Which of the three possible substrates that you tested were used by an enzyme to produce starch?

Question 2

Why were the blank tests necessary?

Question 3

Suggest why **starch phosphorylase** is so named.

3.4 Applications of metabolic control in agriculture and medicine

Metabolic control in agriculture and medicine is widely achieved by the use of enzyme inhibitors which affect metabolism within one particular type of cell or a limited range of plants or animals. The doctor prescribing an antibiotic to treat an infection, the farmer using poison to control rats, or the gardener using an insecticide to kill greenfly on roses are all likely to be using a selective inhibitor.

The biochemist designing a drug or pesticide may achieve selectivity in three ways:

- selectivity due to enzyme differences between organisms
- selectivity due to structural differences between the cells of organisms
- selectivity due to differences in distribution between different types of cell.

Insecticides

All cultivated crops are attacked by insects, ranging from the highly specific carrot fly to the all devouring locust. Chemical methods of insect control have been sought for centuries, but the discovery of organophosphorus compounds was a major breakthrough in insect pest control. These compounds are toxic because they inhibit an enzyme, acetylcholinesterase. Lack of the enzyme disrupts the transfer of impulses from one nerve cell to another. The active site of the enzyme contains an amino acid, which is essential for the enzyme's activity and which reacts covalently with organophosphorus compounds. When this happens the enzyme ceases to function and the resulting disruption of the nervous system kills the insect.

The problem is to find compounds which are toxic to insects but harmless to birds and mammals. Dimethoate is readily absorbed by plant cells and is therefore particularly effective at killing sucking and biting insects. Other insects, which are essential for pollination, are unaffected provided they do not damage plants. Dimethoate is also relatively non-toxic to birds and mammals as they have enzyme systems which rapidly convert it into a harmless compound.

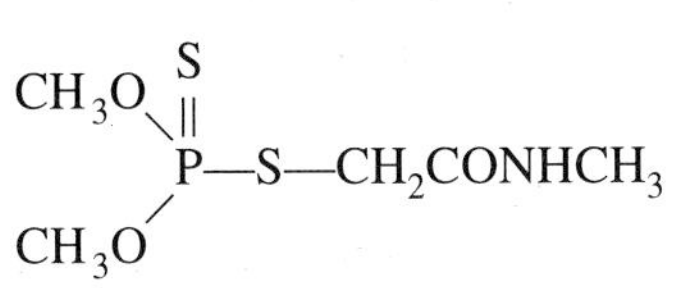

dimethoate

Antibiotics

The antibiotic penicillin was discovered by Alexander Fleming in 1928 but not used until World War II. With Chain and Florey, who developed methods for large scale production, Fleming was awarded the Nobel prize in 1945.

Many bacteria have a strong outer cell wall as well as an inner membrane. An example is *Streptococcus*, which can cause tonsillitis and some forms of meningitis. Penicillin can prevent the formation of the outer wall of bacteria by inhibiting an enzyme involved in the synthesis of the outer wall. Without the protection of their outer wall, they burst when exposed to the conditions present in an animal cell. Human cells do not have this type of cell wall outside their cell membrane and are therefore unaffected by penicillin.

An increasing number of bacteria have become resistant to the range of antibiotics currently available. It is becoming more of a challenge to develop new drugs which will selectively kill bacteria but have minimal side effects in humans and animals. Exposure of bacteria to antibiotics tends to select for resistant types. This is one of the reasons for some people's resistance to the addition of antibiotics to animal feed to prevent bacterial infection under intensive rearing conditions.

Herbicides

Plants depend on photosynthesis for survival whereas animals obtain their energy from food. Triazine herbicides, when transported to a plant's leaves, disrupt the production of ATP from ADP in sunlight; without ATP the plant dies. Examples of triazines are simazine and atrazine. An example of triazine is simazine.

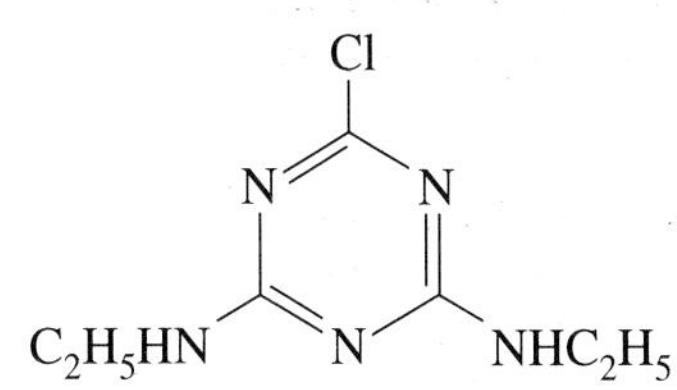

simazine

Some crops, such as sugar cane, are tolerant to triazines because they have an enzyme that can convert the herbicides to non-toxic compounds. The gene for this enzyme has now been transferred to other crops so that weeds can be killed while the crop is growing. But at least 40 species of weed have become resistant to these herbicides.

REVIEW TASK

When you make your own notes on this chapter you should pay particular attention to the following points:

- the meaning of the terms metabolism and metabolic pathway, digestion, and phosphorylation, both substrate level and oxidative
- the interpretation of simple diagrams of pathways (when provided with an outline only), for example the oxidation of glucose, the interconversion of amino acids, sugars and fats, the formation of haem
- the roles of acetyl-CoA, ATP and NAD in metabolism
- the general principles involved in the control of metabolic pathways
- the general principles of selective inhibition
- the selection of suitable examples to illustrate the above points.

CHAPTER 4 NUCLEIC ACIDS

So far we have considered the action of enzymes and we have looked at metabolism, but much of the most exciting research in biochemistry over the past 50 years has involved genes and genetics.

Every living organism carries genetic material. It directs the development of the organism and it is passed on from generation to generation. It must be duplicated precisely every time a cell divides. In this chapter we will seek answers to the following questions:

- what is the structure of the genetic material?
- how is the genetic material copied during cell division?
- how is the genetic information stored and how is the information used to control the activities of cells?
- what happens when there are faults in genetic material?

4.1 DNA, the information molecule

DNA, or **de**oxyribo**n**ucleic **a**cid, was first isolated over a hundred years ago by a Swiss biochemist, Frederich Miescher. He was studying white blood cells obtained from the pus on the bandages of patients recovering after operations. A white precipitate was obtained and found to contain the elements carbon, hydrogen, oxygen, nitrogen and phosphorus. It came from the nucleus of the cells and subsequent experiments showed it to be acidic; so it was given the name 'nucleic acid'.

By 1900 chemical analysis had shown that nucleic acids have very large molecules containing just six components, a sugar, a phosphate group and four different organic bases. However, because of their apparently simple composition, nucleic acids were largely discounted as a possible genetic material. Protein seemed to be a much more likely candidate since it was assumed that the genetic material must be exceptionally complex.

Further research revealed that nucleic acids are polymers made up of monomers called **nucleotides**. Nucleotides contain a sugar, an organic base containing nitrogen and up to three phosphate groups.

The sugar component in DNA is always **deoxyribose**. Deoxyribose is a pentose, which is a five-carbon sugar. Most pentoses have five oxygen atoms but deoxyribose has only four (see figure 4.1).

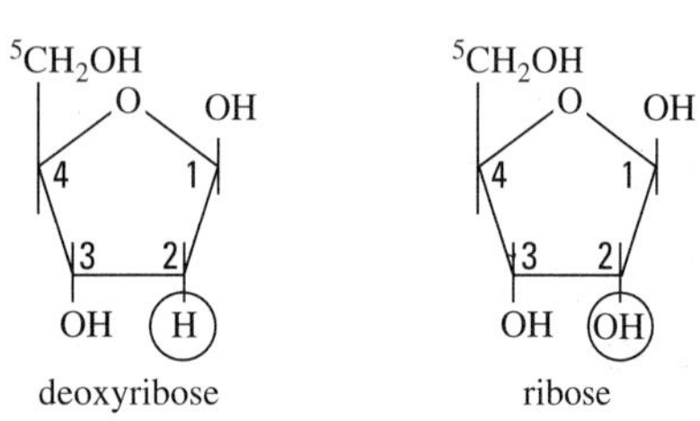

Figure 4.1 Deoxyribose, the sugar found in DNA and the related sugar ribose. Ribose is a component of RNA (see section 4.3).

There are four bases. Two of the bases, **adenine (A)** and **guanine (G),** are purines; they both contain a planar double ring system. The other two bases, **cytosine (C)** and **thymine (T)**, are pyrimidines; they too contain a planar ring system, but with only six atoms in the ring. Sugar molecules link to the (N) atoms (see figure 4.2).

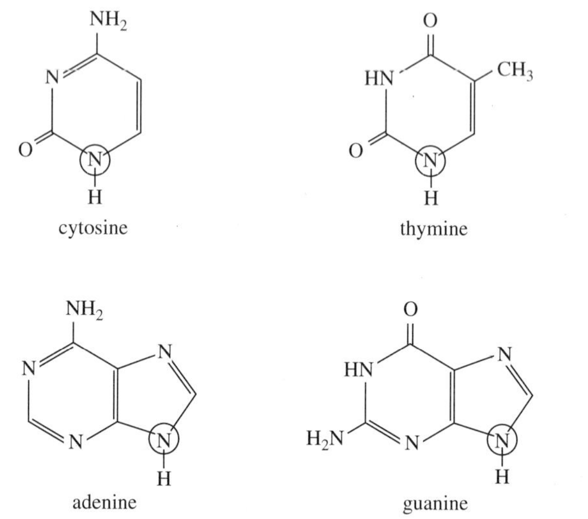

Figure 4.2 The bases found in nucleic acids. In DNA sugar molecules link to the (N) atoms.

A complete nucleotide consists of the three components linked together by covalent bonds. In the diagram of a nucleotide (figure 4.3) notice that the phosphate group is linked to carbon 5′ of the sugar and the organic base to carbon 1′.

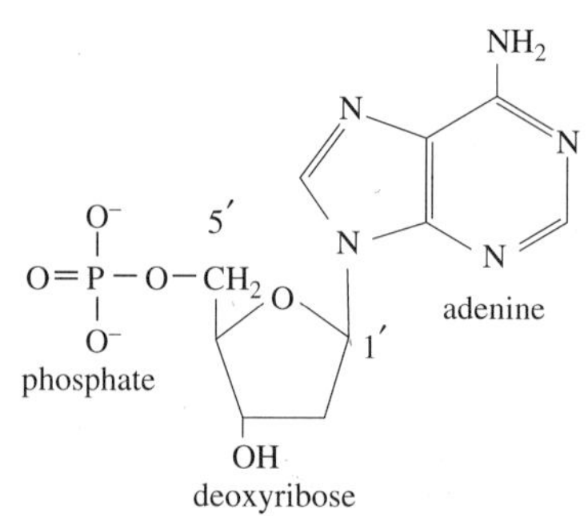

Figure 4.3 A nucleotide (deoxyadenosine monophosphate). Notice that the phosphate group is linked to carbon 5′ of the sugar and the organic base to carbon 1′.

In nucleic acids the nucleotides are linked together in very long chains. The phosphate groups are linked to carbon 3′ and carbon 5′ of each sugar molecule. This gives rise to a long polymer chain of alternate phosphate and deoxyribose units, with organic bases attached to each of the sugar molecules at carbon 1′ (see figure 4.4).

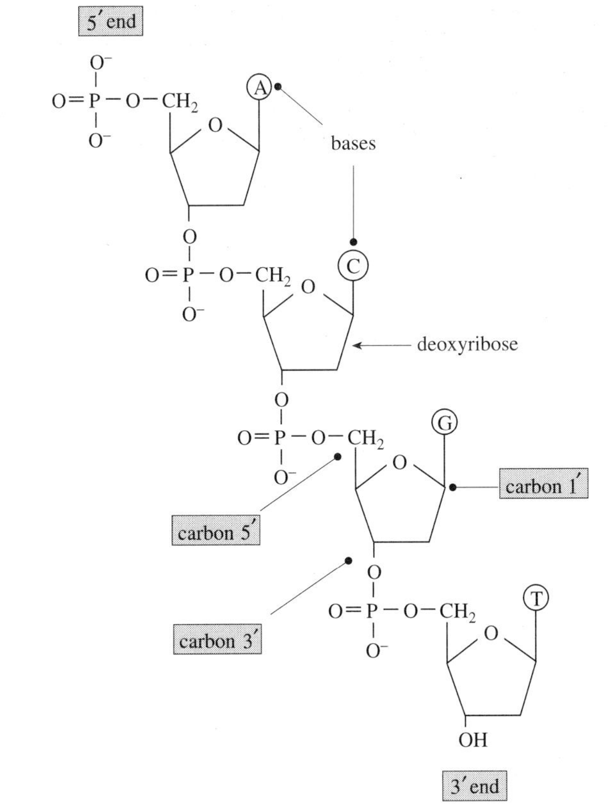

Figure 4.4 The structure of a single DNA chain. Notice that the phosphate groups link to carbons 3′ and 5′ in the sugar.

Important evidence about the composition of DNA was published in 1951 by Chargaff, an Austrian chemist working in the USA. He had determined the relative molar amounts of the four bases in the DNA from many different species and a remarkable feature became apparent. In bacteria, in plants and in animals he found that the number of moles of adenine in the DNA always equalled the number of moles of thymine while the number of moles of cytosine was always the same as the moles of guanine.

Source of DNA	Adenine /molar %	Thymine /molar %	Cytosine /molar %	Guanine /molar %
Bacteria	15.1	14.6	35.4	34.9
Wheat	27.3	27.1	22.8	22.7
Salmon	29.7	29.1	20.4	20.8
Human	30.9	29.4	19.8	19.9

In 1952 Alfred Hersey and Martha Chase, two Americans, carried out experiments which showed for certain that DNA, and not protein, was the genetic material. They were studying a virus that is made of only DNA and protein.

They investigated the infection of bacteria by the virus; they knew that the virus remained on the surface of the bacterial cell but injected genetic material into the cell. After infection of a bacterium, new viral particles are formed and multiply within the host cell. Eventually the host cell is killed and the new viral particles are released.

In one sample of the virus, Hersey and Chase labelled the protein with radioactive sulphur (DNA contains no sulphur). They labelled the DNA in another sample with radioactive phosphorus (protein contains no phosphate). When the bacteria were infected, Hersey and Chase found that the bacteria contained most of the radioactive phosphorus and not the radioactive sulphur. This evidence finally persuaded biochemists that DNA is the genetic material rather than protein.

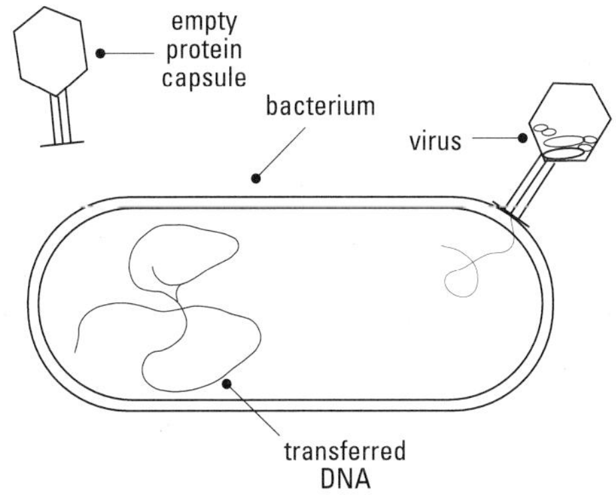

Figure 4.5 The infection of a bacterium by a virus. The DNA of the virus is transferred to the bacterium cell but none of the protein capsule is transferred.

In parallel with the biochemical research, X-ray crystallographers were learning how to obtain good photographs of fibre specimens of DNA. Among the most successful were Maurice Wilkins and Rosalind Franklin, working at King's College, London. It was Franklin's diffraction photographs which generated the more useful data related to the three-dimensional shape of the DNA molecule.

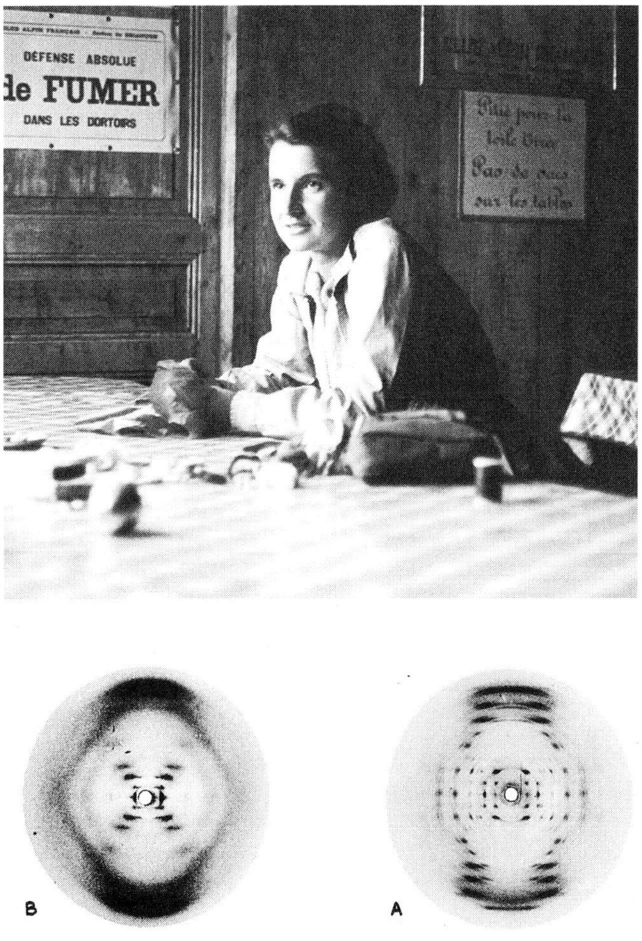

Figure 4.6 Rosalind Franklin and two of her X-ray diffraction photographs of DNA.

By now all the evidence needed to solve the problem was available and scientists in America and Britain were racing to be the first people to publish a paper describing the three-dimensional structure of DNA. Francis Crick and James Watson, working at the Cavendish laboratory at Cambridge, used model building in an attempt to find a structure which was consistent with the covalent structure of DNA, Chargaff's work on the bases and the information from X-ray analysis.

Complementary base pairs	
Purines	**Pyrimidines**
Adenine (A) –	Thymine (T)
Guanine (G) –	Cytosine (C)

Figure 4.7 Francis Crick and James Watson with their model of DNA. In 1962, together with Maurice Wilkins, they were awarded the Nobel prize (Rosalind Franklin had died in 1958).

On 25 April, 1953, Watson and Crick published a paper in *Nature* on the structure of DNA. The article began with a considerable understatement:

'This structure has novel features which are of considerable biological interest.'

They proposed:

- DNA contains two polynucleotide strands
- the strands run anti-parallel
- the deoxyribose-phosphate chains are on the outside
- the bases link the two strands
- the bases are linked by hydrogen bonds
- adenine always hydrogen bonds with thymine, cytosine with guanine
- the two strands are twisted to give a helix.

The structure of DNA is illustrated in figure 4.8. The two strands run in opposite directions (anti-parallel). Look at the deoxyribose molecules at the bottom of the diagram. One has a free OH group on carbon 3′; and the chain runs from 3′ to 5′ up the diagram. On the other strand there is a phosphate group on carbon 5′ and the chain runs from 5′ to 3′ up the diagram.

The bases, in the centre of the diagram, are linked by hydrogen bonds: **A** and **T** by two bonds, and **C** and **G** by three bonds. **A** always pairs with **T**, while **C** always pairs with **G**, and they are called **base pairs**. You can see this more clearly in figure 4.9. Because the bases on one strand always link to their specific matching base on the other strand, the two strands of the helix are known as **complementary**.

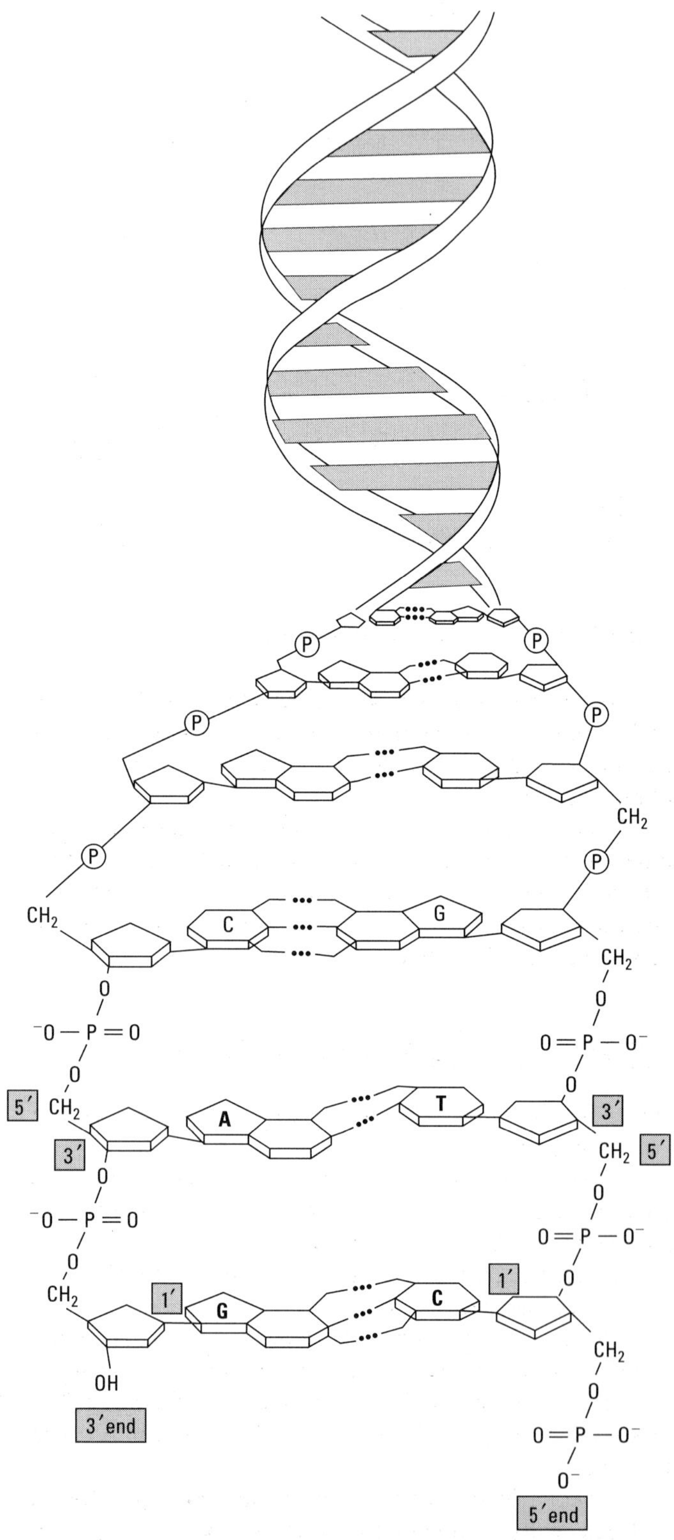

Figure 4.8 The structure of DNA.

In the helical structure, at the top of figure 4.8, the planar base molecules stack one above another like a pile of plates. This maximizes the van der Waals interactions between bases. Indeed the van der Waals interactions actually make a greater contribution to the stability of the DNA structure than the hydrogen bonds between base pairs.

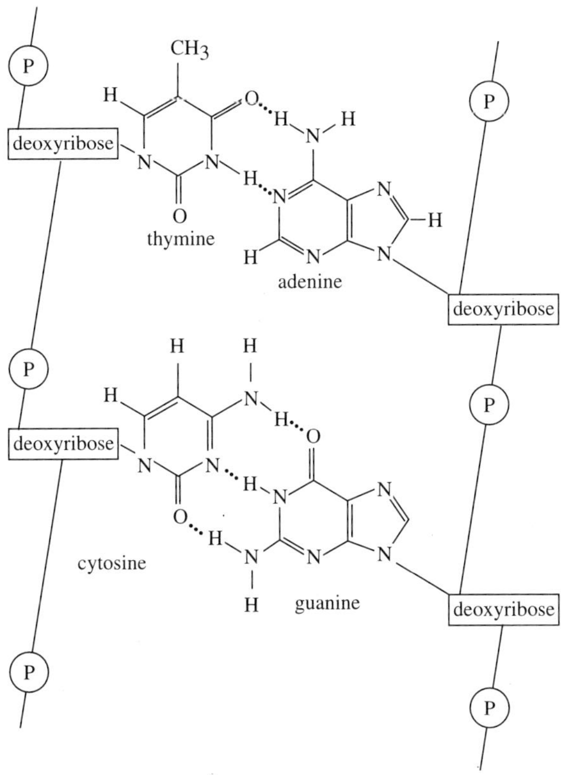

Figure 4.9 Hydrogen bonding between base pairs.

We are now so familiar with the structure of DNA, that we easily underestimate the importance of the discovery. It was perhaps one of the most significant moments in the development of biochemistry.

Since the sequence of bases on one strand determines the order of bases on the other, the Watson and Crick structure offered an immediate and elegant answer to the second question posed at the beginning of the chapter: how is the genetic material copied during cell division?

4.2 The replication of DNA

When Watson and Crick published their model of DNA in *Nature* they wrote

'It has not escaped our notice that the specific pairing we have postulated immediately suggests a possible copying mechanism for the genetic material.'

In cell nuclei, the DNA is packed into relatively large structures called **chromosomes**.

Whenever a cell divides both of the new cells need a complete copy of the DNA in the chromosomes as it contains the information needed for all the activities of the cell. The process of copying the DNA in a cell when it divides is called **replication**.

Nucleic acid replication is a remarkable process with each molecule acting as a template for the synthesis of copies of itself.

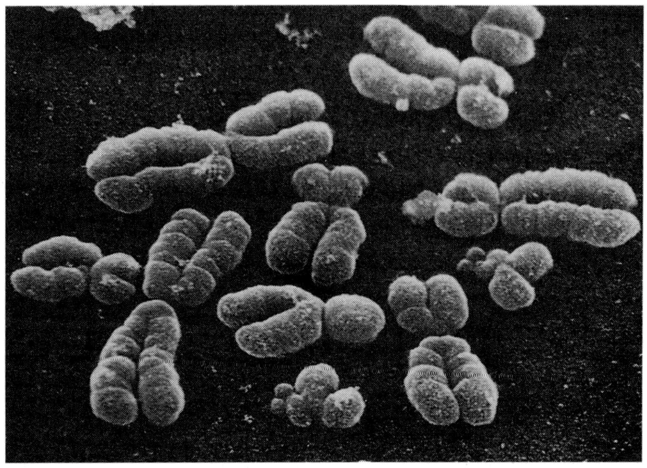

Figure 4.10 Human chromosomes.

During replication, the hydrogen bonds and van der Waals interactions between the base pairs in the double helix are broken and two new strands are formed using the original strands as templates for the synthesis. A new strand contains a complementary sequence of bases as this is dictated by the order of bases in the original strand. Hydrogen bonds and van der Waals interactions then form between original and new strands, creating the stable helical structure.

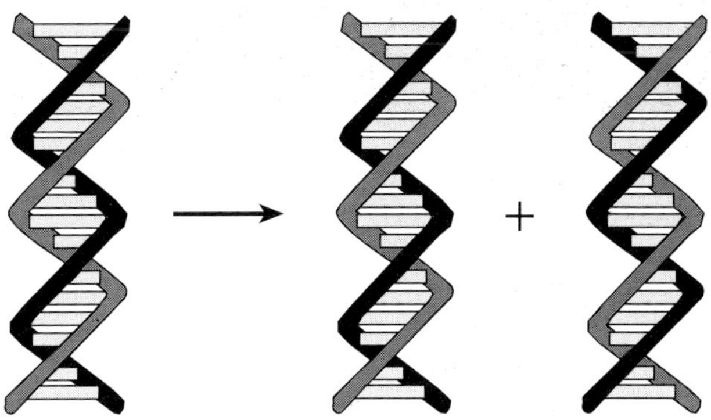

Figure 4.11 The replication of DNA produces two molecules, each containing one strand from the original molecule.

We are now ready to answer our third question, how is information stored in the genetic material and how can it direct protein synthesis?

COMMENT Chromosomes

We have 23 pairs of chromosomes in each of our cells and a single chromosome may have as many as 2000 genes. The complete genetic material is known as the human **genome**. Our DNA is thought to contain about 3×10^9 base pairs and the mass of this DNA is about three picograms, 10^{-12} g. This may not seem very much but if this DNA was unravelled it would be some 2 metres long! Since the nucleus of a cell is about 5×10^{-6} metres in diameter, it is clear that the DNA must be folded extremely compactly in the chromosomes.

The DNA in other species varies from a few thousand base pairs (bp) for a virus to billions in a flowering plant.

COMMENT Replication

The actual process of replication has been found to be very complicated. Firstly the DNA helix is not unravelled from one end to another; this would take far too long (even though this is a fast reaction). The unravelling of the double helix and the polymerization reaction occur simultaneously at a series of replication origins along the DNA chain. A whole series of enzymes and proteins are involved in opening the helix and separating the strands (helicase enzyme), and then holding the strands apart while replication takes place. The fork where a parent DNA divides into two daughter molecules can be seen on the electron micrograph.

Secondly, the new strand of DNA is only made in one direction. The enzyme involved, DNA polymerase, can only synthesize DNA in a $5' \rightarrow 3'$ direction. Since the DNA strands run in opposite directions, one strand can be made continuously as the DNA unravels but the other strand cannot. It is made in short sections, working backwards, which are then linked together by another enzyme, DNA ligase.

Replication requires a large number of enzymes and other compounds. Yet despite its complexity, the accuracy of the replication process is quite remarkable. It has been estimated that the error rate is about 1 mistake in every 10^8 base pairs. There are enzymes which check for any mismatches in the base pairings. The enzyme compares the base sequence on one strand with the complementary sequence on the other strand; once an error has been detected, the faulty section is cut out and replaced. Sections up to 30 bases in length can be repaired.

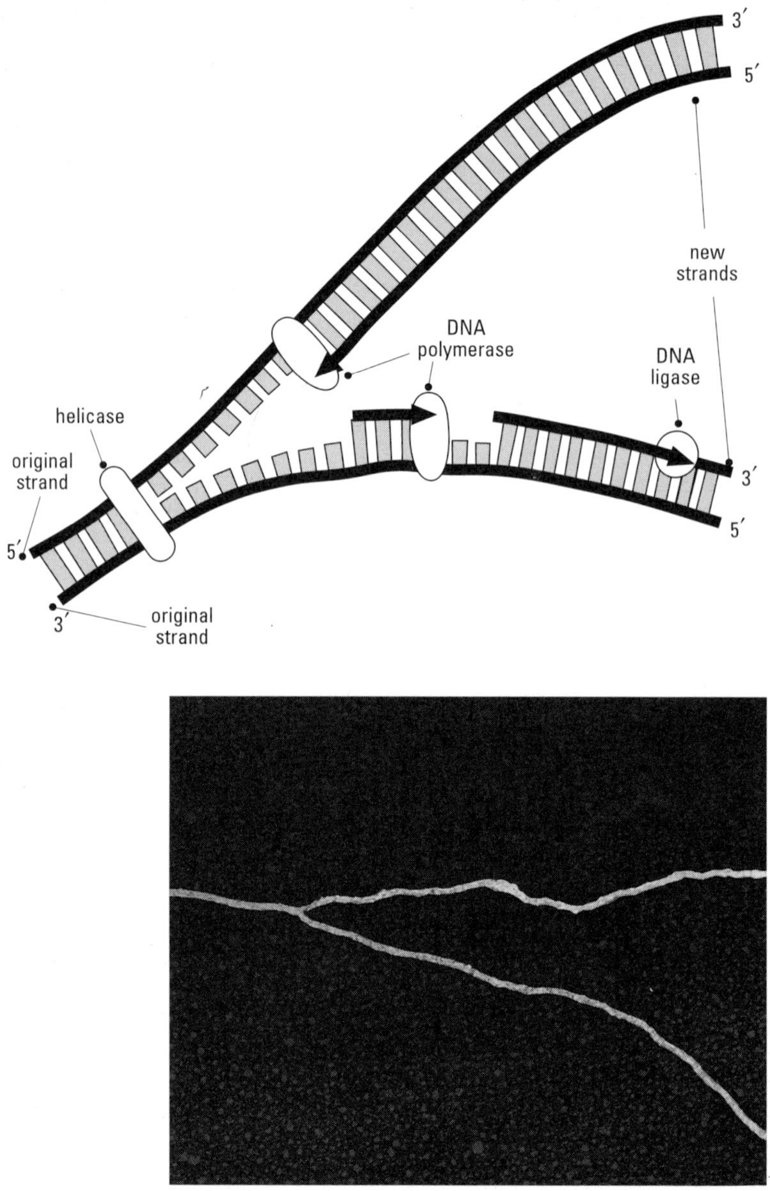

Figure 4.12 DNA replication and forking.

STUDY TASK Replication

In 1958 Meselson and Stahl provided experimental proof of Watson and Crick's hypothesis of replication. Bacteria were grown in a medium in which the only source of nitrogen was ^{15}N. After several generations the bacteria were found to have DNA in which the molecule was heavier than normal. The bacteria were then transferred to a growing medium containing ^{14}N as the only nitrogen source.

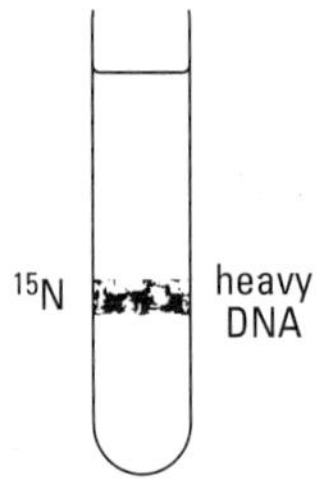

DNA was extracted from the bacterial cells and centrifuged for nearly 2 days in caesium chloride solution. In these conditions the solution develops a density gradient down the centrifuge tube. The DNA from the ^{15}N growth medium is shown in the diagram.

1. How do ^{15}N and ^{14}N atoms differ?
2. Use figure 4.9 to work out the difference in molar mass between a set of base pairs with ^{14}N and a set of base pairs with ^{15}N.
3. After just one cell division in ^{14}N growth medium, a sample of cells had their DNA extracted and centrifuged. Draw a diagram to show the result you would expect.
4. After just one more cell division in ^{14}N growth medium, another sample of cells had their DNA extracted and centrifuged. Draw a diagram to show the result you would expect.
5. What result would you expect after about another 20 cell divisions?

4.3 The synthesis of proteins

A cell might be likened to a computer-controlled machine tool. To make anything useful the machine tool needs access to the computer coded instructions which guide its operations. DNA contains within its base sequence the digital coded instructions which control protein synthesis and thus direct the activities of the cell. But how can a sequence of bases along a strand of DNA direct protein synthesis?

A coded instruction is needed for each of the 20 amino acids that occur in proteins. The four bases on their own could only be used to code for four amino acids, one each. A pair of bases is still not sufficient as it can only generate 16 codes – AG, GA, AT, TA, AC, CA and so on. A sequence of three bases has $4 \times 4 \times 4 = 64$ possible combinations. By using sequences of three bases there are sufficient codes for all the amino acids. Therefore, protein might be coded for by successive sets of three bases along the DNA chain, each triplet coding for an amino acid, with the sequence of triplets determining the order in which the amino acids are to be joined together.

Combinations of the four bases	Number of codes
individual bases	$4^1 = 4$
pairs of bases	$4^2 = 16$
triplets of base	$4^3 = 64$
quartets of bases	$4^4 = 256$

Research has indeed shown that the genetic code is based on triplets. All 64 possible codes are used in DNA. Three of the triplets indicate the termination of a protein sequence, another triplet is a start code (although confusingly the same triplet also codes for an amino acid). The remaining 60 triplets each code for amino acids. This results in nearly all the amino acids having more than one code; for example there are four DNA triplets which code for valine, CAA, CAG, CAT, CAC. Look at these codes carefully and you will see that in this case the first two bases identify the amino acid and the third base just completes the triplet.

Most remarkably the same code is used by virtually all organisms; the *DNA code is universal.* Whether the DNA is from plants or animals or bacteria, the base–amino acid code does not seem to change. Many biologists regard this as the strongest available evidence for the theory that all forms of life have evolved from a common beginning.

The length of DNA which carries the instructions for a particular protein (or part of a protein) is a **gene**. The DNA in a human cell contains sufficient base pairs to code for about three million proteins. However, only about 60 000 proteins are produced from genes. Some DNA is required for control regions, so that proteins are only synthesized where and when required. Other regions of DNA are needed for the production of special RNA molecules. However, much of the DNA has no known use; it may be 'junk' or it may have uses of which we, as yet, know nothing.

Transcription

Protein synthesis does not take place in the nucleus of the cells where DNA is found but on the ribosomes (see figure 2.1). Part of the DNA double helix unravels and a complementary copy of a gene is synthesized (see figure 4.13). The copy diffuses out of the nucleus and moves to the ribosome where protein synthesis takes place. This molecule is called **messenger RNA** (mRNA). It has three of the four bases present in DNA but thymine (**T**) is replaced by a different base, **uracil** (**U**). The sugar is **ribose** instead of deoxyribose.

The mRNA will carry a start code and from that point each triplet of bases along the chain codes for one specific amino acid. A triplet of three bases on the mRNA is known as a **codon**. The complete **genetic code** is listed in figure 4.14, see page 30.

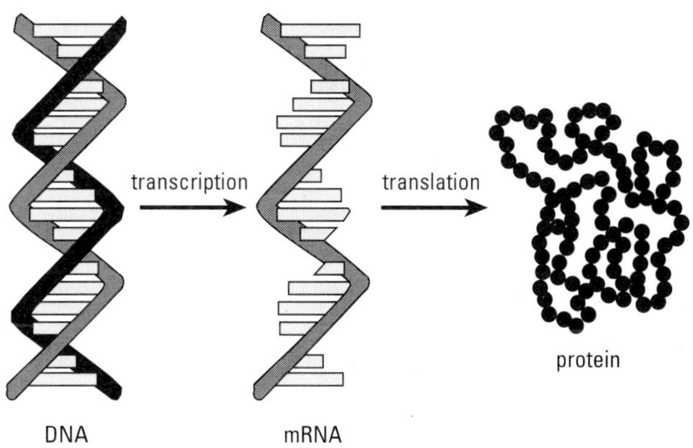

Figure 4.13 Transcription and translation, from DNA to a protein.

Translation

The mRNA at the ribosome directs the bringing together of amino acids in the correct order to produce proteins; this is **translation**. The language of DNA, the order of bases along a strand, is being translated into the language of proteins, an order of amino acids along a chain.

Amino acids on their own cannot bind to mRNA. However, in the cytoplasm, there are small RNA molecules, known as **transfer RNAs** (tRNA). Each tRNA has a specific triplet of bases, an **anticodon**, at one end of the molecule and it can bind the amino acid, corresponding to the complementary codon, at the other end.

Codon base on mRNA	Anticodon base on tRNA
A	U
C	G
G	C
U	A

In the translation process, one codon of the mRNA is paired to its anticodon on a tRNA (see figure 4.15). The correct amino acids are therefore lined up in the correct order. Enzymes link the new amino acids to the growing protein chain through peptide bonds. The tRNA molecules then diffuse away without their amino acids. When a STOP codon is reached synthesis is complete and the protein is released. The protein chain will have folded spontaneously, but at this stage not always to its final tertiary shape.

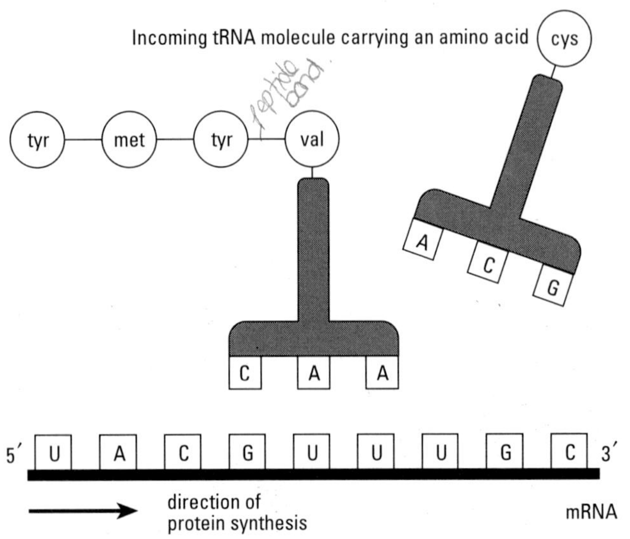

Figure 4.15 Translation. In this simplified diagram the ribosome is not shown.

STUDY TASK The genetic code

1. Look at the genetic code in figure 4.14. How many amino acids have one unique triplet code? How many have two codes? How many have three codes? How many have four codes? Do any have more than four codes?
2. What changes to the third base in a STOP code would produce a code for an amino acid? Which amino acids?
3. Consider a base sequence 5′–UUU AGC CUG AAA GAU CGC–3′ on mRNA. What would be the set of anticodons on tRNA molecules corresponding to this sequence?
4. What would be the amino acid chain corresponding to the mRNA? Use the genetic code in figure 4.14.
5. Assume a single **U** base is omitted from the first codon during transcription. Write out the new base sequence on mRNA and divide into codons.
6. Which amino acids would be linked together this time? What problem has occurred?

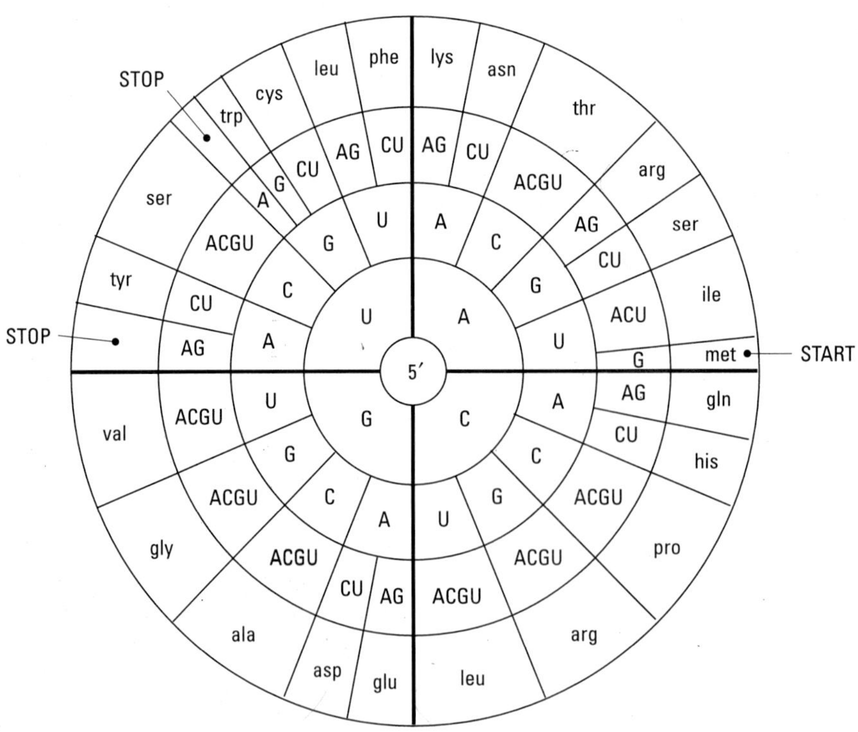

Figure 4.14 The genetic code. Codes are read in the 5′–3′ direction on mRNA and should be read from the centre of the diagram.

COMMENT Transcription and Translation

These two diagrams will give you a better idea of the complexities of the processes of transcription and translation. In the diagram (top) of transcription notice that the bases in DNA are changed to their complementary bases in mRNA, but **U** appears instead of **T**.

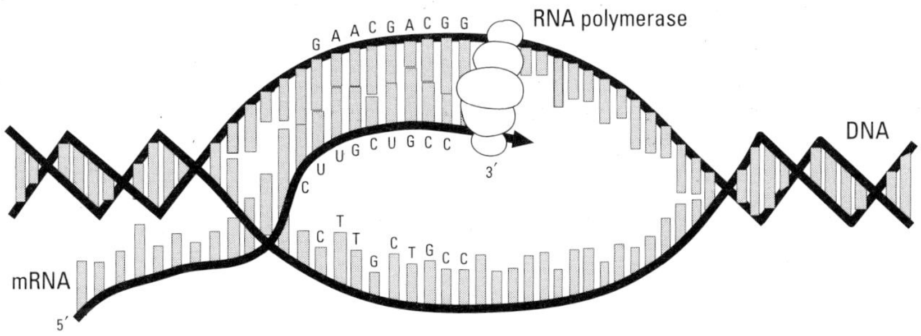

In the diagram (right) notice that several ribosomes can be simultaneously involved in protein synthesis on the same mRNA molecule. You should be able to work out that the anticodons on the tRNA molecules are complementary to the codons they bind to at the ribosome. The tRNA for **asp** is shown diffusing towards the ribosome; **cys** is about to be joined to **val** by a peptide bond: the tRNA for **tyr** is shown diffusing away.

Use the genetic code in figure 4.14 to check the relationship of the codons on the mRNA to the amino acids forming the peptide chain.

Figure 4.16 shows mRNA molecules being made simultaneously along a length of DNA. The molecules grow from short to long 'bristles' as they are synthesized.

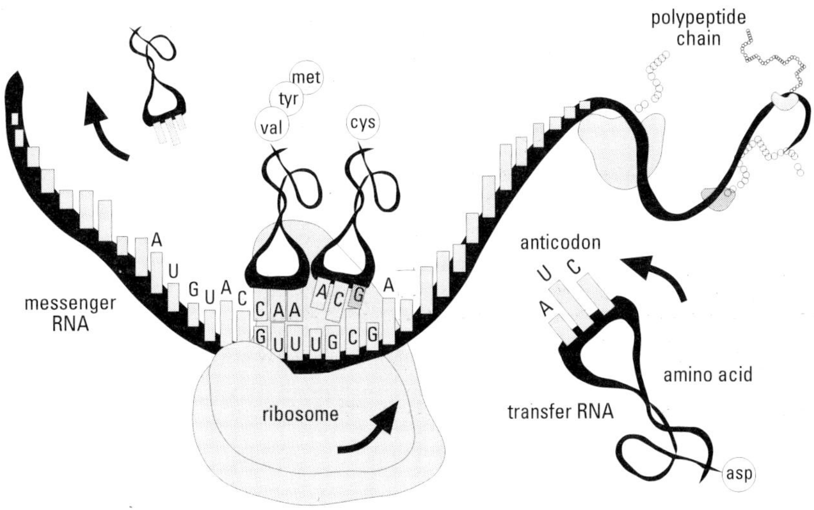

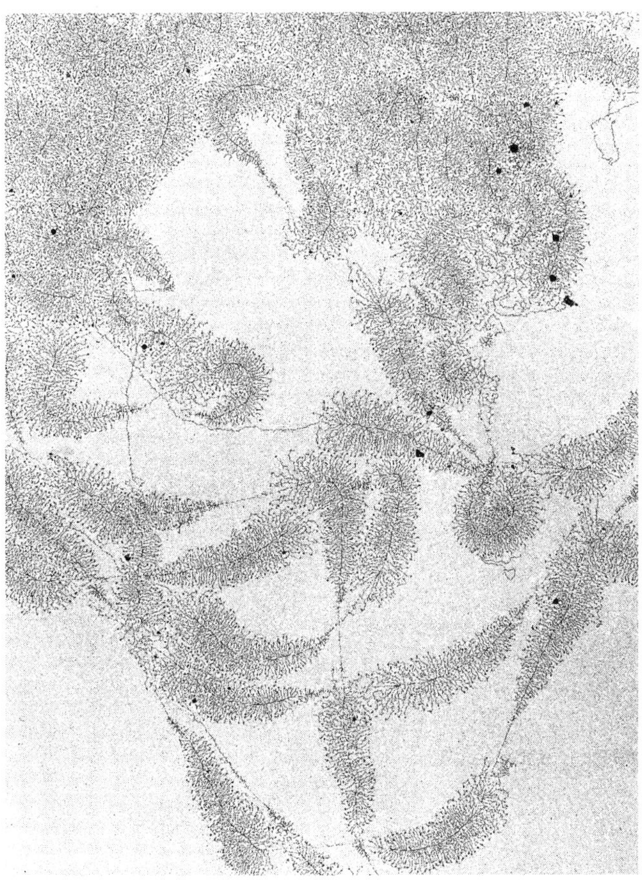

Figure 4.16 Transcription of mRNA from DNA.

4.4 Mutations

Only on very rare occasions during DNA replication is an error not corrected by the cell's own mechanisms. Changes from the original DNA are known as **mutations**. Apart from errors in the replication process mutations can be caused by any process that damages DNA. UV light, cigarette smoke and many other chemical compounds can cause mutations. The chemicals do not have to be synthetic; there are naturally occurring chemicals in many foods, even potatoes, that *can* cause mutations.

In some cases the change in the DNA may be very small; a single base may be miscopied and a single base pair may then be altered in the DNA molecule in future generations. Such mutations are not uncommon.

Since changes in the sequence of base pairs will alter the amino acid coding, the ultimate result may be a change in the structure and behaviour of a protein. In many cases a single change in the base sequence will have no effect on the protein that is being produced. You will remember that most amino acids have several codes and a change in the DNA from a **CAA** sequence to a **CAG** for instance will still produce a protein containing a valine in the correct place. Mutations which produce or remove a START or STOP codon on the other hand may have serious consequences as a critical protein may not be produced or may be so changed that it is unable to fulfil its function.

There is still much to be discovered. The human genome project seeks to sequence all the bases in human DNA. Sequencing of a yeast genome was completed in 1996. Some questions are still unanswered. How can DNA be available for transcription when the chromosomes are so tightly packed? Does the non-coding DNA have a function or is much of it junk? How did such complicated processes evolve?

Much of the research has important technological applications; the diagnosis of disease, the treatment of genetic conditions, and genetic engineering to produce organisms with genes from other species. Some of these issues are considered in Chapter 5.

STUDY TASK The sickle-cell mutation

Sickle-cell anaemia is an inherited disease. Sickle-cell blood cells have an abnormal amino acid sequence in the protein chain of β-haemoglobin. The abnormality is limited to a single amino acid at the sixth position of the 146 amino acid chain:

Normal β-chain	Val	His	Leu	Thr	Pro	**Glu**	Glu	...
Sickle cell β-chain	Val	His	Leu	Thr	Pro	**Val**	Glu	...

Use the genetic code in figure 4.14 to help you answer these questions.

1 What are the possible codes for glutamic acid?
2 What are the possible codes for valine?
3 What base change has occurred in the mRNA for sickle-cell anaemia?
4 What chemical difference *in properties* is there in the side chains of glutamic acid and valine? Use a textbook to find out their structures.

MEDICAL CASE STUDY Cystic fibrosis

Cystic fibrosis is the commonest genetic disorder. It occurs in 1 in 2000 live births and 1 in 22 Caucasians are carriers of the gene. The disease affects the lungs, pancreas, gut and sweat glands. Instead of the normal fluid secretions a thick sticky mucous forms. This viscous mucous blocks and damages the intestines and lungs. Because the supply of digestive enzymes from the pancreas is blocked, nutrients cannot be absorbed and babies fail to thrive, they have repeated chest infections and in particular can get intestinal obstruction. The malfunctioning of the sweat glands results in abnormally salty sweat, which is used as part of the diagnosis of the condition.

The protein coded for by the CFTR gene controls a chloride channel in the membrane of cells lining the airways in the lungs, and surface cells elsewhere in the body. The failure of the chloride channel to open when required means chloride cannot get through to the outer cell surface. One result is that there is less excretion of water to keep surfaces moist and fluid.

The gene is on chromosome 7 and the most common mutation is the absence of codon 508, which codes for phenylalanine. One form of treatment involves delivery to the surface cells of a normal CFTR gene using as a carrier a harmless virus. The virus has been genetically engineered to carry normal human CFTR gene. The virus goes preferentially to surface cells. Developing a gene therapy such as this heralds a new era of treatment.

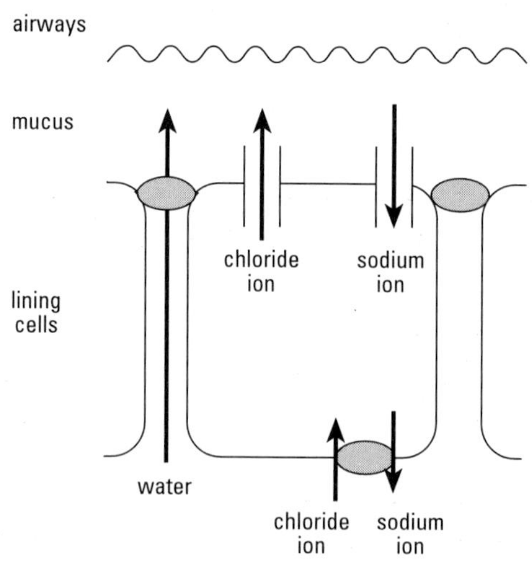

Other treatments involve giving capsules containing high doses of enzyme to help the digestion of fats, antibiotics for chest infections and special drugs to stimulate chloride secretion. The prognosis for patients has improved in recent decades and gene therapy should lead to continued improvements.

4.5 DNA fingerprinting

A DNA 'fingerprint' requires a small sample of your blood, as a source of DNA, and not an impression of the pattern on your fingertips. But it is becoming a more useful and reliable method of identification. It has been possible to confirm the identity of bones found in a mine shaft at Ekaterinburg in Siberia as those of Tsar Nicholas II and his family 80 years after their deaths during the Russian revolution.

The technique was developed by Professor Alec Jeffreys at Leicester University in 1984 and has been publicly available since 1987.

The method depends on some of the 'junk' lengths of DNA between genes. These lengths can vary dramatically between individuals. There are core sequences of about 16 bases that are repeated many times over and are so different that only identical twins will have the same 'genetic markers'.

DNA from a blood sample is treated in a strictly controlled procedure using gel electrophoresis to produce a pattern of DNA fragments based on either a probe which binds only to a specific base sequence or a 'multi-locus' probe which binds to many different types of fragments, producing what is known as a DNA fingerprint. The multi-locus probe produces a much more detailed pattern but the single-locus probe is more sensitive, needing as little as 10 nanograms of DNA.

The procedure is used for the identification of individuals or establishing if there is a family relationship between individuals. It was DNA fingerprinting that established that we are not related to Neanderthals (see Chapter 1).

DNA from a blood sample of a child and his/her alleged parents will confirm whether they are the mother and father with a probability stated to be 99.99% (see figure 4.17).

The method is widely used by the police: the first conviction solely on the evidence of a DNA fingerprint being a Bristol man in November 1987. As the sensitivity of the method has improved the Bristol police have re-examined evidence from 17 cases, going back to 1970 before the method was available.

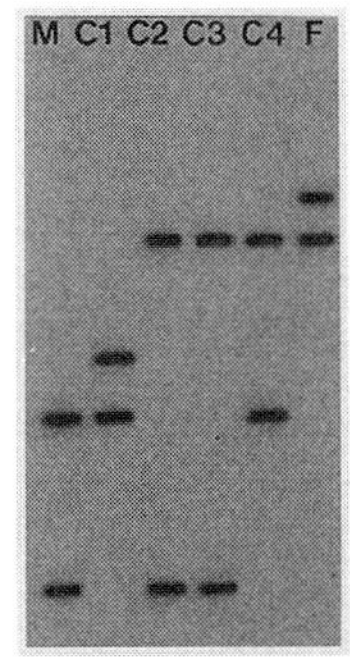

Figure 4.17 Gel electrophoresis results. Since a child inherits half of its DNA from each parent, each band in the child's pattern must be present in either its mother's or father's DNA. In this example F is the father of C2, C3 and C4, but not C1.

One method by which the sensitivity of the method has been improved is the use of the polymerase chain reaction (PCR). In PCR the reaction conditions are arranged so that very small amounts of DNA replicate in a cycle repeated 20–30 times, with the potential to double the amount of DNA at each cycle.

DNA fingerprinting can be used to match an alleged stolen bird to feathers left behind in its nest or cage; badger blood has been used to convict four Yorkshire men accused of badger baiting; and a man, who wrote a letter trying to blackmail a supermarket, was convicted as a result of a DNA test on saliva left on the back of the postage stamp he licked.

Another use of gel electrophoresis is in screening for inherited genetic disorders. This is discussed in the next chapter.

4.6 Extraction of DNA from a plant

DNA can be obtained from any plant or animal source that consists of cells with a nucleus. Fresh cress is suggested for this experiment as a matter of convenience and simplicity. Yeast, onions or peas can also be used but in slightly more complex procedures.

Experiment 4.6 Isolation of DNA from cress

The aim is to isolate nucleic acids from cress tissue. Nucleic acids prepared in this method are sufficiently pure to cut up with restriction enzymes (see next section).

The method uses the detergent **SDS** (**s**odium **d**odecyl **s**ulphate) to disrupt cell membranes and the tertiary structure of proteins, and thus liberate nucleic acids – DNA and RNA – from plant tissue. The detergent solution is pH buffered to minimize degradation of the nucleic acids present. The cells are broken up by gentle grinding. Centrifugation is used to separate cell debris from the nucleic acids in solution. Finally, the nucleic acids are precipitated by adding ethanol.

Procedure

1. Use scissors to harvest about a mortar-full of green cress leaves (about 3–5 g of fresh material). Immediately grind the plant tissue to a smooth purée using a little silver sand and 5 cm^3 of SDS extraction buffer pre-warmed to 65 °C.
2. Pour the ground tissue into a small conical flask, rinsing any material off the pestle and mortar into the flask with a further 5 cm^3 of buffer solution. Loosely seal the top of the flask with aluminium foil and maintain at 65 °C for 30–60 minutes in a water bath. Heat will tend to denature enzymes that would degrade DNA. From time to time, swirl the contents of the flask. Because of the presence of plant pigments the extract may change from green to a brown or purple colour.
3. At the end of the heating period remove the plant debris by centrifuging at about 1000 *g* for 5 minutes, or until a pellet of plant material forms at the bottom of the tube. Take care to balance the centrifuge. The pellet is not needed.

   Note that with slower centrifuges, you may need to extend the time spent centrifuging two- or even three-fold (2500 rpm with a rotor arm of 15 cm corresponds to 1000 *g*).
4. Without disturbing the pellet, carefully pipette or pour off the supernatant liquid into two fresh tubes. Precipitate the nucleic acids (DNA and RNA) by pouring an equal volume of ethanol onto the surface of the liquid extract. A cotton wool-like precipitate of nucleic acids will form over about 10 minutes. Rock the tube gently but thoroughly to mix the two liquids to precipitate the DNA from the plant extract.
5. Collect the nucleic acids by centrifuging at 1000 *g* for 5 minutes, or whatever is necessary. Discard the supernatant liquid and leave the pellet to drain by inverting the tube above a paper towel for 2 minutes.
6. Stopper and label your test-tube. Your sample of DNA is suitable for gel electrophoresis if the necessary apparatus is available.

Question 1

What other water soluble compounds are likely to be present in the supernatant liquid after centrifuging?

Question 2

Why is the effect of detergents on proteins helpful in obtaining a good yield of nucleic acids?

Question 3

Suggest a reason why ethanol precipitates nucleic acids from aqueous solution.

COMMENT GEL ELECTROPHORESIS

The experimental study of samples of DNA and their sequence of base-pairs is based on a technique called gel electrophoresis. But first the DNA has to be cut up into fragments of manageable length. This is done by using one of a group of enzymes that catalyse the digestion of DNA. They are found in bacteria and are called **restriction enzymes** because they are able to restrict the growth and replication of an invading virus by destroying the DNA of the virus.

They cut DNA between two particular bases (*) within a definite sequence of bases.

5′-G * GATCC-3′	5′-GGGCC * C-3′
BamHI	Apa I

After digestion of the DNA sample with a restriction enzyme, **gel electrophoresis** is used to separate the DNA fragments produced. The technique depends on the movement of negatively charged DNA fragments through an electrically conducting gel.

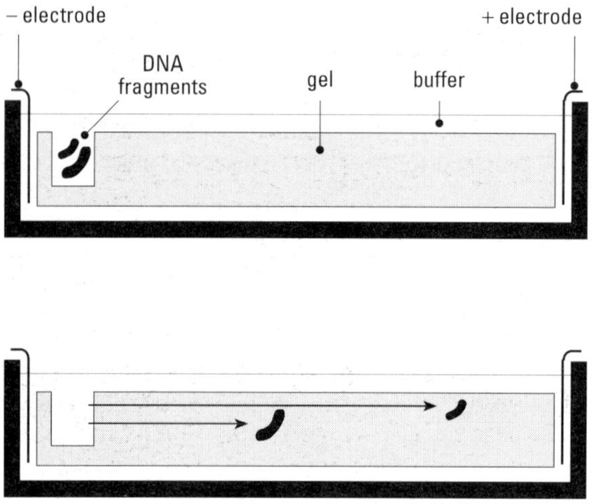

small DNA fragments move further through gel than large fragments

Gels are usually made from **agarose**, a polysaccharide seaweed extract disolved in an alkaline buffer. The gel consists of a sponge-like network of polymer molecules stabilized by hydrogen bonds. When a voltage is applied smaller molecules, or fragments, move faster through the gel than the larger ones so they are separated according to their different sizes.

Small wells in the gel act as reservoirs for the samples of DNA. In order to observe the progress of the electrophoresis a dye, the loading dye, is usually added to each sample.

When electrophoresis is complete, another dye, the staining dye, has to be added to reveal the positions of the invisible DNA fragments.

The diagram shows some typical results.

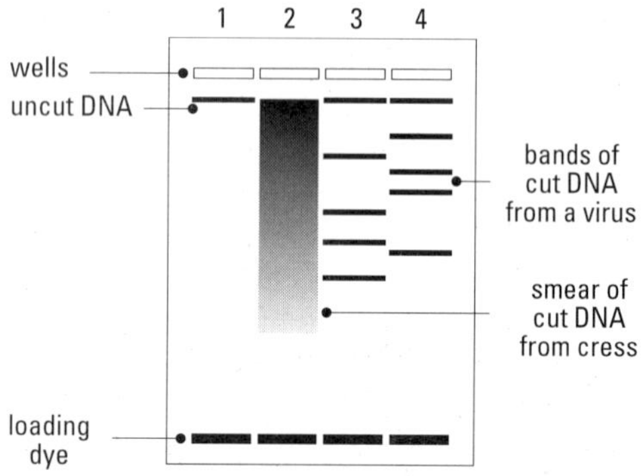

- The strong bands near the wells are uncut DNA.
- The smear down the gel is typical of plant DNA.
- A set of bands is characteristic of a low molar mass DNA.
- The loading dye usually moves further than all the DNA fragments.

REVIEW TASK

At the end of this chapter you should write notes and draw outline sketches to cover the following points:

- the main structural features of the DNA molecule, including the base pairings by hydrogen bonds
- the meaning of the terms nucleotide, gene, genetic code, codon and anticodon, replication, transcription, translation, and mutation
- an outline of the processes of replication, transcription and translation including the roles of mRNA and tRNA
- the relationship of the order of bases in mRNA to the order of amino acids in protein fragments (given the genetic code)
- the interpretation of the results of gel electrophoresis.

You are not expected to recall the structural formulae of the compounds in DNA, nor the abbreviations and names of amino acids.

BACKGROUND READING

The determination of the structure of DNA by Watson and Crick is one of the greatest scientific discoveries of the twentieth century. *The Double Helix* by James Watson, describes the two years of work which led to their discovery. He captures the extraordinary excitement of their desperate efforts to beat other scientists to the solution to one of the great enigmas of the life sciences. *What Mad Pursuit* by Francis Crick is an autobiography and includes a description of how scientists tackled the riddle of the genetic code.

CHAPTER 5 OPPORTUNITIES AND CHALLENGES

Figure 5.1 Beer-making is a tradition that goes back a long time. Above: an eighteenth century brewery, right: modern computer-controlled brewing.

Figure 5.2 A plant with a gene transferred from a firefly: it glows when watered with a compound called lucigenin.

Why are we unconcerned about changes in a brewery's method of producing beer but worried by the transfer of genes from one species to another?

In this final chapter we look at recent developments in biotechnology and, by comparison with traditional methods, indicate some of the concerns people have about genetic engineering.

Micro-organisms

We have always been selecting micro-organisms for our benefit – to produce better bread, or brew better beer. An example is

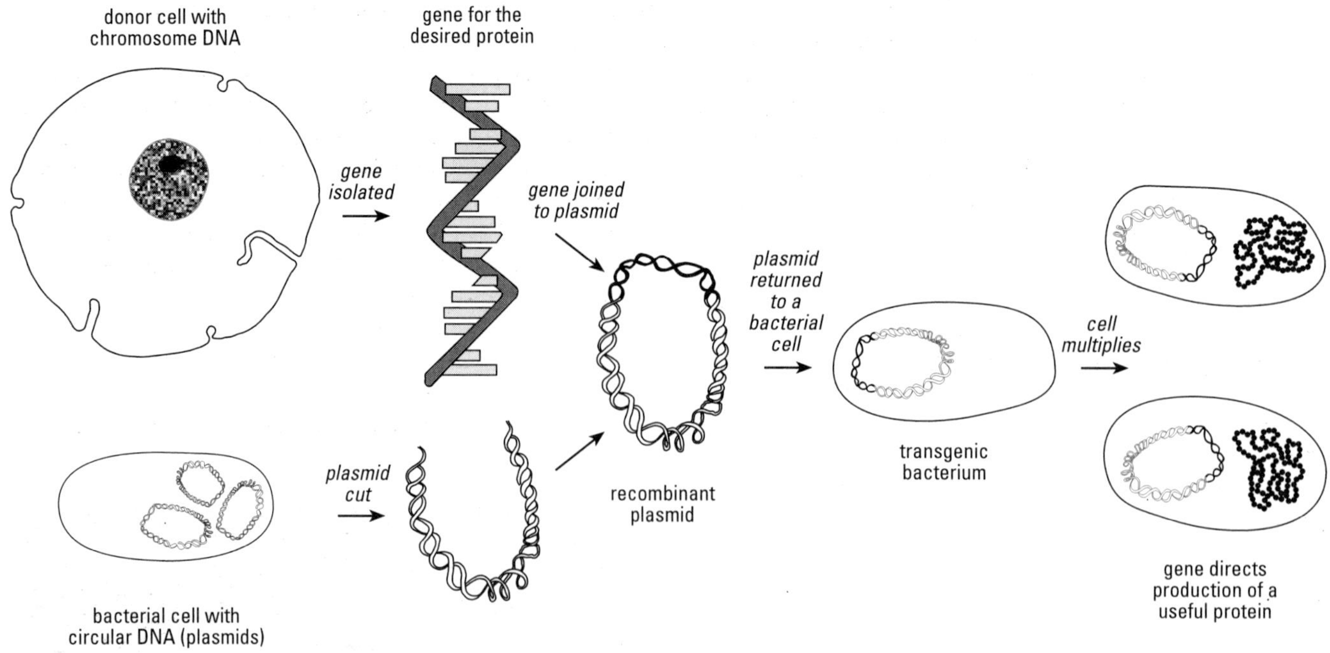

Figure 5.3 How to produce a transgenic micro-organism.

the special strain of *Aspergillus niger* selected for use in the manufacture of citric acid. But the use of micro-organisms that include a gene from a different species is new. A gene, that has been identified as coding for a useful protein, is isolated from a plant or animal. It is then introduced into the DNA of a micro-organism, creating a **transgenic** micro-organism (see figure 5.3). The transgenic micro-organism can then be grown in large quantities and the useful protein extracted as a by-product of its growth.

An example is the production of human insulin using either the bacterium *E. coli* or a yeast. The gene for human insulin, in a modified form, has been transferred to the DNA of yeast. The genetically engineered yeast is fermented with glucose in 80 000 dm^3 tanks for three weeks. A precursor form of insulin is synthesized on the ribosomes in the yeast cells and then folds into the correct tertiary structure. The insulin is excreted by the yeast cells into the fermentation broth and can be isolated by centrifugation and absorption. Further treatment removes a peptide connecting the two chains of insulin and produces active human insulin.

Genetically engineered insulin for diabetics has been available in Europe since 1982. Previously insulin had to be obtained from slaughtered cows and pigs, whose insulin differs in a few amino acids from the human form.

Another example is the treatment of children with pituitary dwarfism who do not grow to a normal height. However, they can be treated with human growth hormone. The hormone used to be obtained from the pituitary glands of human cadavers, but the supply was very limited. Furthermore, this was not a safe source as was shown in the 1980s when several people died from Creutzfeldt-Jakob disease caused by contamination of the hormone. Growth hormone is now obtained as a product from a genetically modified bacterium. This is a valuable advance but it is not without problems. Difficulties arise when the drug is given to children significantly below normal height, but not suffering from pituitary dwarfism.

Plant breeding

During the last 50 years conventional plant breeding and improvements in agricultural methods have more than doubled the yield per hectare of crops such as wheat, barley, maize and rice. The new technology of genetic engineering involves the transfer of specific genes to a plant, in some cases from a completely different species, a transfer that would never happen naturally. Results can be obtained more rapidly than through conventional plant breeding.

One method of transferring single genes to plants involves the use of a micro-organism that naturally infect some plants. During the infection process this micro-organism transfers part of its DNA into the DNA of the plant. Scientists are able to select single genes and add them to the piece of DNA that transfers. When the micro-organism is allowed to infect plants the selected gene is also transferred. In some of the modified plants that are produced the gene may turn out to be inactive (gene silencing), or unstable (altering in effectiveness over the growing season or over several generations), or not inherited (the gene is not transferred from one generation to the next). The plants also have to be tested to ensure that the plant's own genes are still functioning normally. Companies, that have identified valuable genes to use in transgenic plants, naturally want to protect their investment by patenting the genes and the technology.

One of the plants studied has been oilseed rape. It is the fourth largest commercial crop in the UK and the seed oil is an important raw material for margarines, cooking oils and confectionery. Coconut and palm trees, that grow in the tropical climate of south-east Asia, produce oils containing valuable lauric acid. It is used to make soaps and detergents. The addition of a gene from the California bay tree to oilseed rape has resulted in transgenic plant seeds that contain 40% lauric acid. The acid can therefore now be obtained from plants grown in temperate regions. It has been suggested that the impact on the agricultural industry of developing countries like Malaysia and the Philippines could be devastating. However, it is more likely that in the long run genetic modification of coconut and palm trees to alter the oil they produce will have a major impact on tropical oilseed production. It is important to recognize that novel crops produced by genetic engineering have to be economically competitive in the context of world agriculture.

Crops can be made resistant to specific herbicides by the addition of a gene from a another species. Phosphinothricin is a non-specific herbicide, and a gene from the bacterium *Streptomyces* produces an enzyme that can inactivate it. By transferring the gene into crop plants they can be made resistant to levels of the herbicide that will kill weeds growing among the crop. So fields can be sprayed, killing weeds without affecting the crop.

Three main risks have been suggested as being associated with the growing of agricultural crops of transgenic plants:

- the transferred genes may escape via pollen and be transferred to closely related plants, potentially spreading herbicide resistance
- genetically modified plants may spread into natural communities of plants in an uncontrolled way
- the growing of transgenic plants may alter the balance of micro-organisms in the soil.

In considering these risks we need to recognize that large amounts of random gene transfer occur naturally and is part of the process of evolution. Many are not convinced by this argument. The process of producing genetically modified plants needs to be closely regulated.

COMMENT THE REGULATION OF GENETIC ENGINEERING

In the European Union there are Directives which set legal guidelines for biotechnology research and the use of genetically modified organisms. These Directives are incorporated into UK law by Acts of Parliament or regulations so that companies have to have a licence before genetically modified seeds can be grown as a crop. To help deal with the rapid developments in this area a Food Standards Agency (FSA) is planned.

Policy is being developed on the basis of the reports on topics such as the ethics of including in food products from genetically modified sources (Polkinghorne), the ethics of new technologies in animal breeding (Banner), and a Nuffield Council on Bioethics; and the British Ecological Society has produced guidelines on how to go about risk assessment for genetically modified organisms.

Animal breeding

Traditional animal breeding has resulted, after many centuries, in the varied species of dogs that are so prized today and our farm animals that are all significantly different from those on farms a century ago. But now, as with plants, interest is shifting to developing new characteristics in animals by the use of the techniques of genetic engineering.

The addition of a novel gene to an animal's chromosomes is as well understood as the process in plants: but work with animals is more expensive and the time scale is longer. In practice 98% of work on transgenic animals is done with mice, mainly to study the treatment and diagnosis of human disease.

The aim is to produce a group of animals, all with an extra gene which gives the animals a novel characteristic. In Scotland there are now flocks of sheep that include copies of human genes in their make-up. One gene codes for the enzyme alpha 1-antitrypsin, used in the treatment of hereditary cystic fibrosis. The enzymes are produced by normal protein synthesis in the sheep and can be extracted from their milk. In another flock a gene codes for factor IX, a blood clotting agent used in the treatment of haemophilia B. Obtaining factor IX from the milk of disease-free sheep (see figure 5.4) is likely to prove a more reliable and safer source than human blood from blood donors. At the time of writing this (1998) it will be several years before these treatments are generally available, assuming that clinical trials go well.

Figure 5.4 This lamb, named Polly, has had a human gene added to her chromosomes so that she will produce a human blood clotting agent in her milk.

Other experiments have been directed at increasing animal growth and this has been achieved in fish by the transfer of genes for growth hormones. The farming of such fish needs to be treated with considerable caution, to avoid possible escape into the natural environment, because of the competitive advantage of 'rapid growing' fish over wild species.

Screening for genetic disorders

Genetic screening in adults can be achieved by taking a small sample of blood, using restriction enzymes to cut up the DNA and gel electrophoresis to separate the fragments by size (as described in Chapter 4). Chemiluminescent or radioactive probes, that bind to very specific fragments of DNA, are then used to look for the DNA that carries the mutations responsible for the disorder. Doctors were able to test for over 450 genetic disorders in 1998.

Genetic screening can also be used in early pregnancy to find out if the foetus has inherited a disorder from its parents, such as sickle-cell anaemia but taking samples from a fetus increases the risk of a miscarriage.

There are some conditions such as anencephaly (absence of a brain) and Tay-Sachs disease (resulting in blindness, loss of mobility and death within 5 years) where parents usually elect for a legal abortion when they are told of the problem. Huntington's disease is an example of a more difficult situation:

people with the condition develop involuntary muscular movements and find it more and more difficult to look after themselves. However, Huntington's disease usually only appears in people over the age of 35.

The moral issue for parents is whether to seek an abortion or let a child be born knowing that a distressing condition will develop. The ethical issue for doctors and society is the extent to which genetic screening should be available. New knowledge brings new issues to challenge our sense of moral responsibility.

Why do biochemists want to produce transgenic plants and animals?

Genetic engineering works because the genetic code is universal – a gene will encode the synthesis of the same protein whatever species it is in because of our common ancestry. Genetic engineering has become a topic of considerable public concern, with supermarkets issuing leaflets to inform their customers, and developments in the UK regularly reviewed by advisory committees and tightly regulated by the government.

For biochemists, as for most scientists, the motivation for their work is to solve problems and make discoveries; so the original research into DNA started and will continue 'because it's there'.

For industrial companies, it's rather different. They are interested in finding new products that they can sell, and this means the new product has to be in some way better, safer or cheaper than any alternative product. So industrial companies want to develop products based on genetic engineering because they can see advantages in them and expect to be able to sell them. The most important benefits should be increased crop yield, better nutrition and improved diet. While the World population continues to increase the contribution of genetic engineering to improving our standard of living should be significant.

For the public, other questions seem important. Gene transfer from bacteria to bacteria may be acceptable on the grounds that the exchange of genes occurs in nature (see figure 5.5); and moving genes within a species can be regarded as just speeding up conventional breeding programmes. But the transfer of genes across the normal species barrier occurs rarely, and transfers between plants and animals would never be expected to happen naturally, so to attempt this by genetic engineering is unacceptable to some people. At one level or another people may object that we are 'tampering with nature' and feel that developments should go no further.

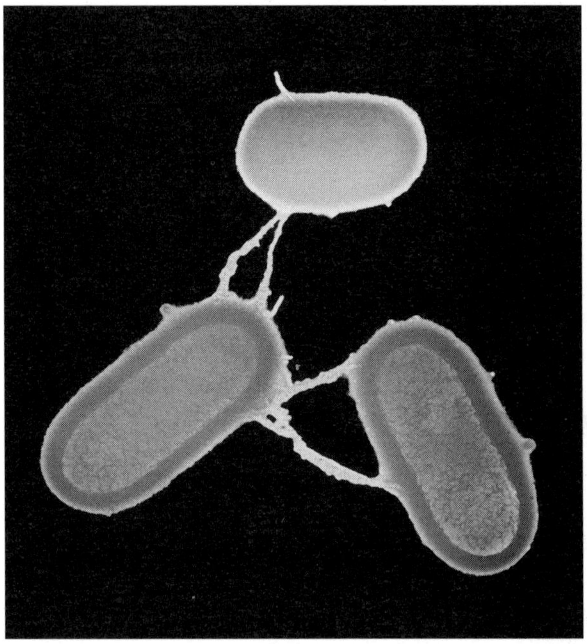

Figure 5.5 Bacteria preparing to transfer DNA in a naturally occurring process.

COMMENT

Conflicting views are expressed about genetic engineering but there are some distinctions which can help to clarify the issues.

Firstly, there is the distinction that can be made between morality and ethics. Moral concerns are considered to be the beliefs that are personal to an individual, while ethical concerns are the standards by which a community regulates its behaviour. For example, vegetarians have decided that it is morally wrong for them to eat meat, but in Britain there is no general objection to eating meat provided certain ethical concerns are respected about the way animals are treated.

Secondly, moral and ethical positions can be based on either fundamental reasons or on pragmatic reasons. Some vegetarians believe that it is fundamentally wrong to kill animals, as Buddhists do, but other people will be vegetarians because they do not agree with modern farming methods and would therefore rather not eat meat. Fundamental reasons are also called intrinsic reasons; pragmatic reasons are called extrinsic.

Pragmatic reasons do not constitute valid objections to a belief based on fundamental reasons; nor should the ethical position of a community prevent a person following their own reasonable moral code.

Shakespeare could not have been thinking of DNA when he wrote in *All's Well that Ends Well*

'The web of our life is of a mingled yarn, good and ill together'

but his words can stand as a metaphor for what we know now about our DNA. Indeed Francis Crick, who *does* know about DNA, has said:

'Our fate is in our genes.'

REVIEW TASK

Write a short account of the main applications of biochemistry, drawing on the preceding chapters in this book as well as this chapter. Make a list of the main types of application and for each type identify an important benefit and also a potential risk.

GLOSSARY

ACTIVE SITE The place in an enzyme where the catalytic reaction occurs.

ADP (adenosine diphosphate) A nucleotide with two phosphate groups; it forms when ATP is hydrolysed in an energetically favourable reaction.

AGAROSE GEL A network of carbohydrate polymers in water that restricts the passage of protein and DNA molecules, depending on their size.

ALA (amino laevulinic acid) A precursor to haem in a metabolic pathway.

ALLOSTERIC INHIBITOR A compound that reduces the activity of an enzyme by binding at a specific site, that is not the active site.

AMPLIFICATION A reaction process that increases the number of copies of a sequence of DNA (see PCR).

AMYLASE An enzyme in saliva that can catalyse the hydrolysis of starch.

ANTIBIOTICS Compounds that kill or prevent the growth of micro-organisms, such as bacteria

ANTICODON The sequence of bases on tRNA that is complementary to a codon on mRNA.

ATP (adenosine triphosphate) A nucleotide with three phosphate groups; formed from ADP in an energetically unfavourable reaction.

BACTERIA Single cell micro-organisms that can reproduce themselves; they commonly causes disease.

BARBITURATES Drugs that have a sedative effect on the human central nervous system.

BASE PAIR A pair of bases that bond together in DNA by hydrogen bonds; also link tRNA to mRNA. **C**ytosine pairs to **G**uanine, **A**denine pairs to **T**hymine or **U**racil.

BIOTECHNOLOGY The scientific use of living organisms to produce useful products, on a laboratory or industrial scale.

CATALASE An enzyme in liver that catalyses the decomposition of hydrogen peroxide.

CHROMOSOME A DNA molecule folded into a compact structure; found in the nuclei of cells.

CITRIC ACID CYCLE The pathway by which ethanoyl groups are oxidized to carbon dioxide and water.

CLONES Cells, or organisms, with identical sets of chromosomes; from a single parent cell.

CoA (coenzyme A) An essential substrate in a number of biochemical reactions.

CODON The sequence of three bases on mRNA that codes for one of 20 amino acids; there are also start and stop codons.

COFACTOR A compound or ion that is required for an enzyme to be active.

COMPETITIVE INHIBITOR Binds reversibly to an enzyme, blocking the active site and reducing enzyme activity.

COMPLEMENTARY DNA A matching strand of DNA, with a complementary set of base pairs.

DEHYDROGENASES Enzymes that remove two hydrogens from a molecule.

DENATURE The unravelling of the tertiary shape of a molecule; this is usually reversible in DNA but not in proteins.

DNA (deoxyribonucleic acid) An organic polymer with a chain of phosphate groups and pentose sugars, with side chains of four organic bases; usually found as a double-stranded helix.

DNA FINGERPRINT A unique pattern of DNA fragments obtained by gel electrophoresis; everyone has their own unique pattern.

DNA POLYMERASE The enzyme that catalyses the formation of two double stranded DNA molecules using single strands of DNA as templates.

ELECTROPHORESIS A method of separating charged molecules in a gel or on paper by the application of a voltage.

ENZYME ACTIVITY The rate at which a reaction takes place in the presence of a particular amount of enzyme.

ENZYMES Proteins that catalyse specific biochemical reactions by lowering the activation energy of the reaction.

ESSENTIAL AMINO ACID An amino acid which is needed in the diet because it cannot be synthesized by a particular mammal.

GENES The long sequences of bases in DNA that code for proteins.

GENETIC CODE The complete set of the possible sequences of three bases (codons) that can occur in mRNA and their translation into amino acids or stop and start codes.

GENETIC ENGINEERING The ability to alter the genetic material of an organism in a deliberate manner, in particular moving genes from one species to another.

GENOME The complete DNA in the chromosomes of an organism.

GLYCOGEN A glucose polymer that acts as an energy store in the human body.

GLYCOLYSIS The pathway in which glucose is oxidized to pyruvate.

HALF-LIFE The time for the concentration of a compound to halve from an initial concentration; for biochemical reactions half-lives are usually constant (independent of the initial concentration).

HERBICIDE A compound designed to kill unwanted plants, such as weeds.

HOMOGENIZATION The process of breaking the cell membrane to release the content of the cell.

HORMONES Compounds that help to co-ordinate the activities of different cells.

IMMOBILIZED ENZYMES Enzymes that are bonded to insoluble support materials or trapped inside an insoluble material.

INHIBITION The reduction of the activity of an enzyme by a specific chemical compound.

INSECTICIDE A compound designed to kill harmful insects.

INSULIN One of the hormones that regulate the concentration of glucose in the blood.

IRREVERSIBLE INHIBITION The irreversible binding of a compound to the active site of an enzyme that permanently reduces its activity.

LACTATE The anion of lactic acid, 2-hydroxypropanoic acid.

LIPASE An enzyme which hydrolyses the ester group in fats (lipids).

LYSOZYME An enzyme that can destroy the cell walls of bacteria.

METABOLIC PATHWAY A sequence of reactions in cells.

METABOLISM The reactions that maintain cells and living organisms.

MICROBE A micro-organism that can replicate.

MITOCHONDRIA The part of a cell where the citric acid cycle, oxidative phosphorylation and haem synthesis take place.

mRNA Synthesized using DNA as a template; moves to the ribosome to direct protein synthesis.

MUTATION A change in the sequence of the bases on DNA

NAD^+ (nicotinamide adenine dinucleotide) A compound that is an oxidizing agent in many biochemical reactions.

NUCLEIC ACID A biochemical acid formed from a sugar, organic bases and phosphate groups; see DNA and RNA.

NUCLEOTIDES The monomer building blocks for DNA and RNA; they consist of phosphate groups, a pentose sugar and an organic base.

NUCLEUS The part of a cell where chromosomes (DNA) are found.

OXIDATIVE PHOSPHORYLATION The formation of ATP by addition of a phosphate group to ADP; by a reaction coupled to the oxidation of NADH.

PCR (polymerase chain reaction) A laboratory procedure for obtaining multiple copies of a target length of DNA.

PEPSIN An enzyme secreted in the stomach.

PHOSPHATE The anion of phosphoric acid, H_3PO_4.

PHOSPHORYLATION A reaction that adds a phosphate group to a molecule, such as ADP.

PLASMIDS DNA molecules in which the strands have joined in a circle; capable of self-replication.

PRIMARY STRUCTURE The amino acid sequence of a protein.

PRIMERS Short lengths of DNA (or RNA) that mark the site on large DNA strands where replication will start; used in PCR.

PROBES Short lengths of DNA that will bind to specific sequences of bases in DNA, enabling them to be detected; used in DNA fingerprinting.

PROTEASE An enzyme which hydrolyses the peptide group in proteins.

PURINES Organic bases with two rings, both with nitrogen atoms; adenine and guanine are purines.

PYRIMIDINES Organic bases with two nitrogen atoms in a six-membered ring; cytosine, thymine and uracil are pyrimidines.

PYRUVATE The anion of pyruvic acid, 2-oxopropanoic acid.

RECOMBINANT DNA Technology process of cutting and then rejoining DNA molecules to alter their genetic make-up.

REPLICATION The production of two identical double-stranded DNA molecules from one DNA molecule.

RESTRICTION ENZYME Cuts DNA at a specific sequence of bases.

RIBOSOME The part of a cell where protein synthesis occurs.

RNA (ribonucleic acid) A nucleotide polymer that differs from DNA having uracil instead of thymine, and ribose instead of deoxyribose.

SECONDARY STRUCTURE Regular folding in the amino acid chain of a protein.

SUBSTRATE A compound involved in a chemical reaction, catalysed by an enzyme.

SUBSTRATE LEVEL PHOSPHORYLATION Is the formation of ATP during one of the reactions by which food molecules are broken down, by the transfer of a phosphate group from a substrate to ADP.

TBE BUFFER (tris-borate-EDTA) A buffer that provides the right conditions for the gel electrophoresis of DNA.

TE BUFFER (tris-EDTA) A buffer used when cutting DNA with restriction enzymes.

TERTIARY STRUCTURE The complete three dimensional structure of a protein.

TRANSCRIPTION The formation of mRNA that is a complementary copy of a gene, but with thymine replaced by uracil.

TRANSGENIC ORGANISMS Organisms that have a gene in their DNA from another species, put there by laboratory methods (genetic engineering).

TRANSLATION The production of a protein molecule using the information on a mRNA molecule.

tRNA The group of compounds that carry amino acids to the ribosome for protein synthesis.

VIRUS An infectious agent that can only replicate in a host cell; not usually regarded as a living organism.

EXAMINATION QUESTIONS

Section A

A1

(i) Write the full name of DNA. (1)

(ii) Describe TWO structural features of DNA and indicate how the stability of the structure is maintained (*the precise chemical structures of the various parts of the molecule are not expected*). (4)

(iii) What is meant by a **base pair**? (1)

(iv) What is the significance of the sequence of the organic bases in a strand of DNA? (2)

(v) During transcription a molecule of mRNA is made. Explain the importance of base pairing during transcription. (2)

(vi) State TWO ways in which a mRNA molecule differs from a DNA molecule. (2)

(12 marks)

A2

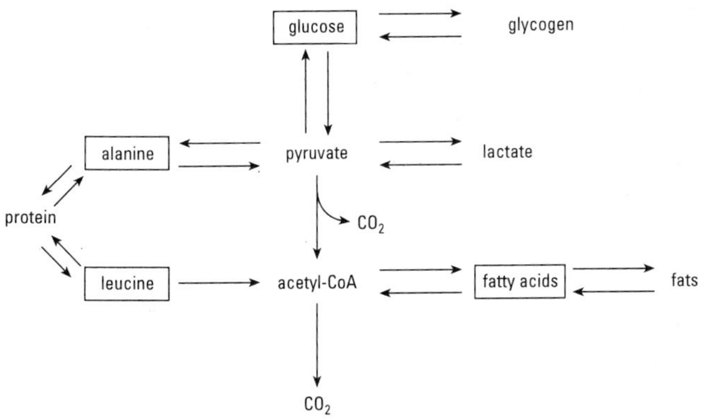

(i) Why is surplus fat in the food supply of a mammal unable to be stored as glycogen? (2)

(ii) What is an essential amino acid? Use the diagram to explain why leucine is an essential amino acid but alanine is not. (4)

(iii) What type of reaction is the conversion of pyruvate to acetyl coenzyme-A? What reagent is required? (2)

(iv) Why is the conversion of ADP to ATP in metabolism of importance to mammals? (2)

(10 marks)

A3 The schematic diagram below shows the formation of mRNA by transcription from DNA.

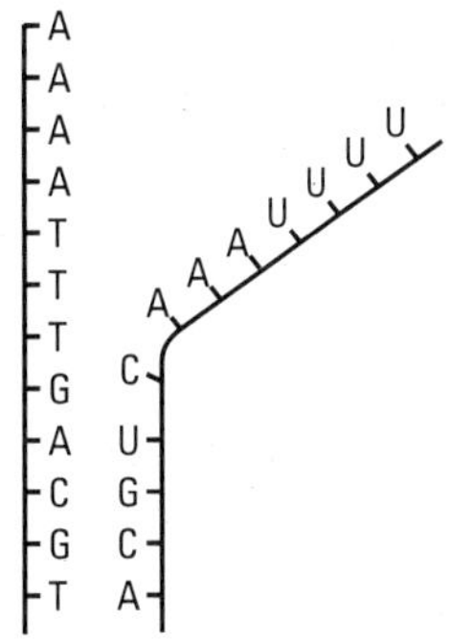

(i) Write the full name of RNA. (1)

(ii) Write the names of the bases denoted by A, C, G, T and U. (3)

(iii) mRNA is involved in the synthesis of proteins. Explain how the order of the bases on mRNA dictates the primary structure of proteins. (3)

(iv) How many amino acids are coded for by the mRNA in the diagram? (1)

(v) How might an error in the order of bases on a DNA molecule affect the metabolism of the cell? (3)

(11 marks)

A4 ATP, adenosine triphosphate, plays a key role in metabolism.

(i) What is the key role of ATP? (1)

(ii) Describe the TWO types of phosphorylation which occur in animal cells and which lead to the synthesis of ATP from ADP. (4)

(iii) Where does the energy come from that is needed for the synthesis of ATP from ADP? (1)

(iv) State THREE types of cellular activity which involve ATP. (3)

(v) In the absence of enzymes, the hydrolysis of ATP to ADP has a very high activation energy. Explain why this is essential in order for ATP to hold such a key role in metabolism. (2)

(11 marks)

A5

(i) Describe briefly the biochemical function in an animal cell of the following:

mitochondria
ribosomes
cell membrane. (3)

(ii) A homogenate can be made from cells by grinding them in buffer solution to release the cell contents.

(a) Why is buffer solution used for this purpose rather than water?

(b) Outline the procedure used to separate the mitochondria, in the homogenate, from other cell material. (3)

(iii) The cytoplasm from cells can be tested for enzyme activity. Sketch a graph of reaction rate against substrate concentration for an enzyme catalysed reaction. Explain what is happening at high substrate concentrations. (3)

(iv) Sketch a graph of reaction rate against pH value of the reaction mixture for an enzyme catalysed reaction. Account for the shape of your graph. (3)

(12 marks)

A6 The enzyme pyruvate decarboxylase catalyses the reaction:

$$CH_3{-}\overset{\overset{\displaystyle O}{\|}}{C}{-}CO_2^- + H^+ \longrightarrow CH_3{-}\overset{\overset{\displaystyle O}{/\!/}}{C}{-}H + CO_2$$

(i) Using a partly purified preparation of the enzyme, outline how you could measure its activity in the above reaction. *Experimental details are not required.* (2)

(ii) Pyruvate decarboxylase requires thiamine pyrophosphate and magnesium cations as **cofactors**. What are cofactors? (2)

(iii) Assuming that pyruvate decarboxylase has the typical properties of an enzyme, describe and explain what changes would be expected to occur to the enzyme molecule when the enzyme was heated to 70 °C for a few minutes. (3)

(iv) The pyruvate binds to the pyruvate decarboxylase at its active site. The active site contains several critical amino acids. What do you understand by the term 'critical amino acids' in this context? (2)

(9 marks)

A7

(i) What are enzyme **inhibitors**? (1)

(ii) Explain clearly what is meant by the following three types of inhibitor and explain how they operate:

irreversible
competitive and
allosteric. (6)

(iii) Describe how enzyme inhibitors can be used in insecticides. (2)

(iv) How are enzyme inhibitors important in the control of metabolic processes? (2)

(v) State, and explain, an example of enzyme inhibition. (3)

(14 marks)

Section B

B1 The effect of pH on the activity of amylase was shown by the following experiment.

Five test-tubes were set up each containing 5 cm^3 of 1% starch solution and 1 cm^3 of 0.2 M sodium chloride solution. 2 cm^3 of buffer solution, of a different pH in each case, was added to each test-tube.

A **suitable volume** of a solution of amylase was added to one test-tube and a clock started. The time taken for the starch to react was determined by testing drops of the solution at 30 second intervals with iodine, using a spotting tile.

The experiment was then repeated with the other test-tubes in turn.

(i) Name the type of reaction in which the starch is used up. To what product is the starch finally converted in this reaction? (2)

(ii) Describe a chemical test for presence of starch. (2)

(iii) Describe the procedure for determining a **suitable volume** of amylase solution for the experiment. (3)

(iv) Explain the possible function of the sodium chloride in these experiments. (3)

(v) Sketch a labelled graph of the results you would expect to be obtained. (2)

(vi) It was suggested by a student that a more effective method for following the reaction would be to add some iodine solution to the reaction mixture in the test-tubes and follow the course of the course of the reaction by using a colorimeter. State, giving reasons, whether this method would be successful. (2)

(14 marks)

B2 The enzyme catalase is found in liver and catalyses the decomposition of hydrogen peroxide solution.

$$2H_2O_2(aq) \longrightarrow 2H_2O(l) + O_2(g)$$

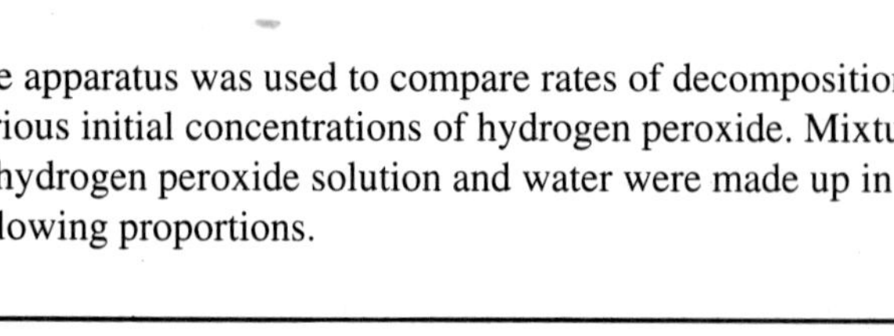

The apparatus was used to compare rates of decomposition at various initial concentrations of hydrogen peroxide. Mixtures of hydrogen peroxide solution and water were made up in the following proportions.

Experiment	1	2	3	4	5	6	7
Volume of hydrogen peroxide / cm^3	0	2.5	5	7.5	10	12.5	15
Volume of water / cm^3	15	12.5	10	7.5	5	2.5	0

In each experiment, the mixture was introduced into a clean flask along with 10 cm^3 of phosphate buffer, pH 7. The plunger of the gas syringe was pushed to zero and the self-sealing cap fitted. 2 cm^3 of catalase solution were then rapidly injected through the self-sealing cap, using a hypodermic syringe. The flask was shaken vigorously, and the volume of gas collected in the gas syringe was noted after 15 seconds.

(i) The same apparatus was used for each of the seven experiments. What precautions should be taken between the experiments? (1)

(ii) Why was buffer solution added to the reaction mixture? (2)

(iii) What was the purpose of Experiment **1**? (2)

(iv) A graph of volume of oxygen produced against the initial hydrogen peroxide concentration was plotted. How would the syringe readings be corrected before the graph was plotted? (2)

(v) Sketch the graph that would be obtained and explain its shape in terms of the mechanisms of enzyme-catalysed reactions. (6)

(vi) If Experiment **7** was repeated, but with enzyme that had been boiled and cooled, what volume of gas would be given off? Explain your answer. (3)

(16 marks)

B3 A red pigment was extracted from the skin of a North African snake. It was found that the liver of the same snake yielded an enzyme which had the effect of catalysing the decomposition of the red pigment to produce a colourless substance. It was found that the amount of pigment varied considerably from snake to snake.

Further investigations suggested that the activity of the enzyme depended on the concentration of magnesium ions, with an optimum concentration of magnesium ions needed for maximum catalysis. This could then account for the variation in pigmentation of the snakes.

(i) Bearing in mind the principles that govern the design of the experiments to investigate enzyme activity, outline an experimental procedure which could test the suggestion that there was an optimum magnesium concentration for the decomposition of the red pigmaent. (10)

Use the following facts as further guides:

- the snake's body temperature is normally 25–30 °C
- the liver tissues of the snake are at a constant pH of 7.6
- 10 cm^3 of an 0.01 M solution of the pigment are decolourized in 5 minutes by 1.0 cm^3 of an 0.0010 M solution of the enzyme
- a 1.0 M solution of magnesium ions inhibits the reaction completely.

(ii) Assuming the suggestion proves to be correct, sketch the shape of the graph for the rate of reaction against magnesium ion concentration. (2)

(12 marks)

Section C

C1 Antibiotics, insecticides and herbicides must be selective to be effective.

(i) State, and explain, ONE example of a compound that is effective because organisms have **different structures**. Outline how the compound works. (4)

(ii) State, and explain, ONE example of a compound that is effective because organisms have **different enzyme systems**. Outline how the compound works. (4)

(iii) Describe how genetic engineering can use resistance to herbicides to improve crop yields. (2)

(10 marks)

C2 Many diseases are inherited. They arise because of mutations.

(i) What is meant by a mutation? Explain how a mutation can result in an ineffective enzyme. (3)

(ii) Name a genetically inherited disease and describe briefly some of the symptoms of the disease. What is the nature of the molecular defect associated with the disease? (4)

(iii) What technique do biochemists use to identify genetic conditions? (3)

(iv) Suggest how genetic engineering might help to treat genetic diseases. (2)

(12 marks)

C3 Attempts have been made to use immobilized enzymes to treat diseases.

(i) What is an immobilized enzyme? (1)

(ii) State, and explain, advantages that an immobilized enzyme might have over one that was not immobilized. (4)

(iii) Describe and explain TWO ways by which an enzyme could be immobilized. (6)

(iv) If an immobilized enzyme was taken by mouth in an attempt to treat a disease, suggest why would it tend to be ineffective. (1)

(12 marks)

C4

(i) What is a hormone? (1)

(ii) Using insulin as an example, explain how hormones can control metabolic processes. (5)

(iii) Outline the production of insulin from a transgenic source. What are the advantages of producing insulin from a transgenic organism rather than an animal source? (5)

(11 marks)

General editor,
Nuffield Advanced Chemistry,
3rd edition
Michael Vokins

Editor, Special Studies
Michael Vokins

Author of this Study
Roger Barnes

Safety adviser
Peter Borrows

This is the Students' book of BIOCHEMISTRY, part of the Nuffield Advanced Chemistry course. Each Study provides opportunities for students to extend their knowledge in a particular field of applied chemistry. Students look at the ways in which chemistry is used in medicine, industry, and agriculture, and at the economic and social responsibilities of the scientist.

This *Students' book* examines the properties and structure of enzymes, metabolic pathways and the structure and function of DNA. Among the applications of Biochemistry studied are genetic engineering, the use of enzymes in industry and applications in agriculture. The treatment is introductory as the course is planned to last for four weeks. Experimental work is included that supports the topics being studied.

ISBN 0-582-31188-8

9 780582 311886

LONGMAN